For Collectors Only®

Swiss Magazine Loading Rifles 1869 to 1958

by

JOE POYER

EDITED BY

CRAIG RIESCH

2nd Edition, Revised

North Cape Publications®, Inc.

The author would like to thank all those who helped, directly or indirectly, with this project, particulary Brian Anderse; Ken Catero; Ed Cote; Jim Gronning, Gruning Precision Gunsmith; Roy Marcot, Old Fort Lowell Armory Press; J. Michael Metzgar and Ken Fladrich, Armory of Orange, Orange, CA.

The author would especially like to acknowledge and thank Eric A. Timrud for his extraordinary efforts in assisting the author in researching the Vetterli and Schmidt rifles and in reviewing and correcting the manuscript. Also, Ian Shein of the Red Star Military Museum and James Wesson for allowing their extensive collections of fine Swiss rifles and carbines to be photographed for this book.

.

ISBN #978-1-882391-32-5
North Cape Publications®, Inc. P.O. Box 1027,
Tustin, California 92781
714 832-3621, Fax 714 832-5302
E-mail: ncape@ix.netcom.com
Internet Website http://www.northcapepubs.com
Printed by Delta Printing Solutions, Valencia, CA 91355

Swiss Magazine Loading Rifles

Table of Contents

Foreword

The "For Collectors Only®" series of books are designed to aid the arms collector in familiarizing him- or herself with a chosen firearm, or line of firearms. Each book is thoroughly researched from (whenever possible, and providing they still exist) original design and production records. Government and military documents, reports and records are scrutinized for information pertaining to their use. And large numbers of examples are examined in detail. North Cape Publications®, Inc. acquires examples of original firearms in good condition for extensive analysis, which always includes complete disassembly and reassembly. During this process, parts are measured and photographed. They are compared to the same parts at different stages during the production life of the firearm to ascertain when changes and modifications were made.

Once the research base has been compiled, it is organized so that it can be presented to the collector in the most efficient fashion. The heart of any "For Collectors Only" book is the "part-by-part" description of the firearm. To this is added a historical description of its design and production. In the case of military firearms, chapters are devoted to the accoutrements and equipment issued directly to the soldier to support and care for it. An exploded or other view of the disassembled firearm and its nomenclature is prepared; a discussion of the ammunition issued for the firearm as well as a section describing the care and maintenance and the assembly and disassembly of the firearm is included. Lastly, a glossary of terms as well as a research bibliography of related material is compiled.

The collector is always the focus of the book, and we strive to organize and present the material in such a way that the collector can easily and quickly analyze a firearm to determine its originality and authenticity right down to the least part.

In this book, the author has attempted to describe the world's first successful general issue military repeating small arms, the Vetterli and Schmidt rifles, developed and manufactured in Switzerland in the late 19th century.

The author has concentrated on those rifles and carbines issued to the Swiss military and police forces as these are the rifles and carbines that are encountered most frequently by North American collectors. Nonmilitary and experimental models are exceptionally scarce outside Switzerland and are rarely encountered on this side of the Atlantic. For those lucky enough to possess one of the nonmilitary or experimental Vetterli or Schmidt rifles, the author suggests consulting *Die Repetiergewehre der Schweiz* by Christian Reinhart, Kurt Sallaz and Michael am Rhyn and published by Verlag Stocker-Schmid, Dietikon-Zürich, Switzerland. While published in German, there are currently several translation aids available on the Internet to non-German speakers.

Finally the author notes that the Schmidt rifle is often referred to in North America as the Schmidt-Rubin rifle. Rudolf Schmidt was the designer of the rifle, and Eduard Rubin developed the cartridge used by the rifle. The rifle, therefore, is generally referred to in this book as the "Schmidt" rifle.

The K31 was a redesign of the Schmidt rifle performed during the 1920s, and many collectors do not consider it a "Schmidt design." But in North America the K31 is generally known as a "Schmidt" rifle and is therefore referred to generically as a "Schmidt" in this book, except where details of its differences are concerned. These are the author's personal preferences, and those who disagree are perfectly welcome to continue referring to the rifle as the "Schmidt-Rubin" and to distinguish the K31 as a wholly separate design.

Conventions

1. Following the editorial practice of North Cape Publications®, Inc., major changes to parts are classified as "Types." For instance, "two kinds of lower barrel bands were developed and used. The first is designated as Type 1 and the second as Type 2," etc. This system is

used for the convenience of collectors only in trying to keep track of changes to parts over the years and was never used by Swiss Ordnance officials.

2. A listing of serial numbers by year is not available for the Vetterli rifles. Because the author was not able to locate sufficient Vetterli rifles to produce a realistic statistical sampling, it was not possible to develop a serial number listing by year.

Serial numbers for the Schmidt rifles were compiled by Christian Reinhart, Kurt Sallaz and Michael am Rhyn for their book, *Die Repetiergewehre der Schweiz, Die Systeme Vetterli und Schmidt-Rubin.* Their tables are used here with permission.

3. All markings on a rifle part are shown in quotation marks, i.e., "*Waffenfabrik.*" The quotation marks were not part of the marking unless so noted.

4. The languages most commonly used in Switzerland are German, French, Italian and Romansch. As German is the dominant language spoken by more than 65% of the population, it is only natural that the markings on the Vetterli and Schmidt rifles are in German.

4. When measurements are given from a point to a hole or other opening, that measurement is understood to be to the center of the hole or opening, unless otherwise noted.

5. All reference directions are given from the shooter's standpoint while looking toward the muzzle with the rifle shouldered. Thus right side refers to the side with the bolt handle, etc.

6. Line drawings of parts may be used in preference to photographs for clarity or to emphasize certain aspects. Occasionally, a drawing or part of a drawing is exaggerated to emphasize a point. Where exaggeration is used it is noted in the text and/or caption.

7. The Vetterli and Schmidt rifles and all their variations were designed and manufactured according to the metric system of measurement. Because this book is written primarily for an English-speaking audience, all dimensions are in decimal inches, feet and yards. For those more familiar with the metric system, decimal inches can easily be converted to millimeters by multiplying by 25.4. Example: 2.3 inches x 25.4 = 58.4 millimeters.

8. Information may be repeated several times in the text at different points so the reader does not have to page back and forth.

9. All firearms shown in this text are from the North Cape Publications®, Inc., collection, unless otherwise credited.

10. The author would appreciate any comments or corrections the reader would care to make. They may be sent in care of North Cape Publications®, Inc.

The photographs of these two unidentified Swiss soldiers in full field uniform on this and the page oppposite were taken in the mid-1890s. Both are in full marching gear and pose with their Model 1889 rifles. Brian Kostel collection.

R. Liebig
Atelier Mercur
ZÜRICH III.
Grüngasse 10.

Swiss Magazine Loading Rifles

Model 1869 Vetterli Infantry Rifle

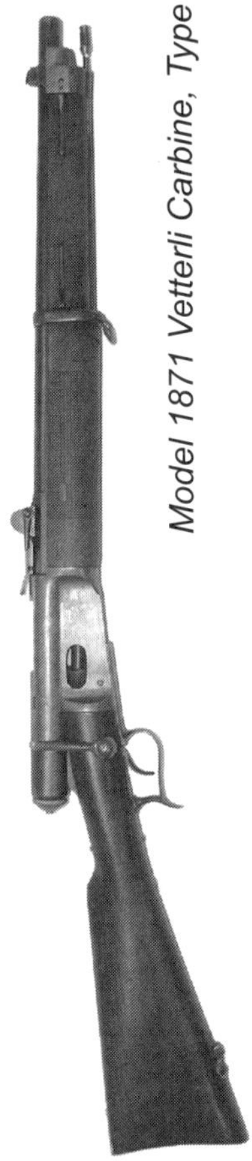

Model 1871 Vetterli Carbine, Type 2

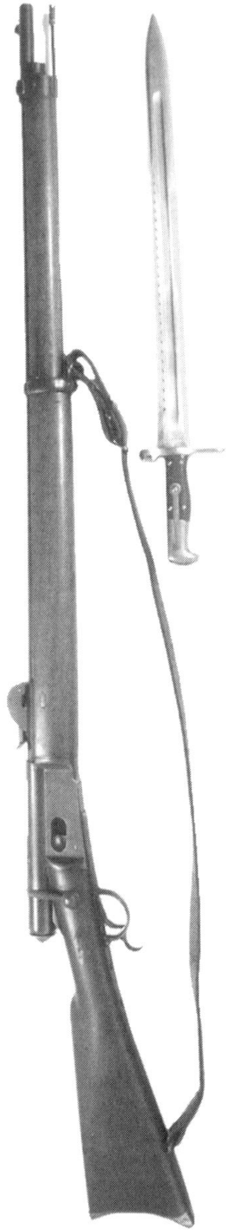

Model 1881 Vetterli Infantry Rifle

***Note**: all rifles and carbines shown on pages xiv and xv are from the North Cape Publications®, Inc., collection.*

Swiss Magazine Loading Rifles

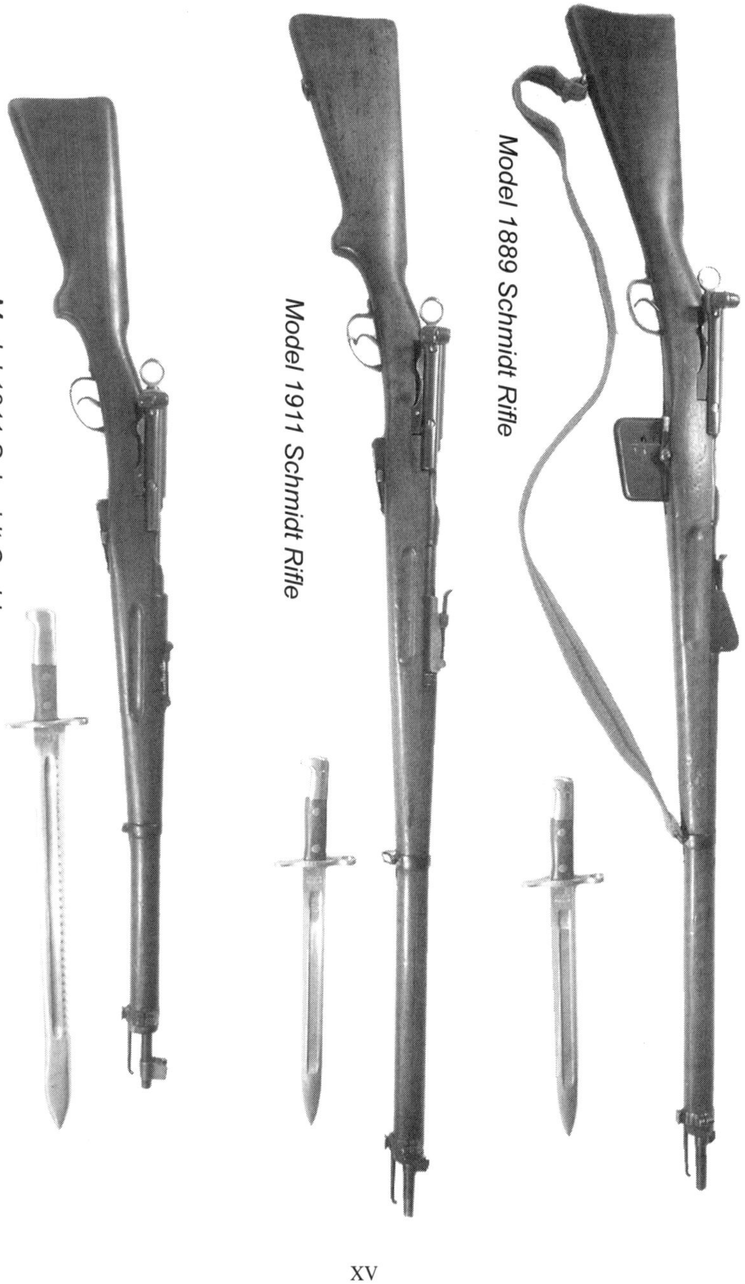

Model 1911 Schmidt Carbine

Model 1911 Schmidt Rifle

Model 1889 Schmidt Rifle

RUBIN MAGAZINE RIFLE

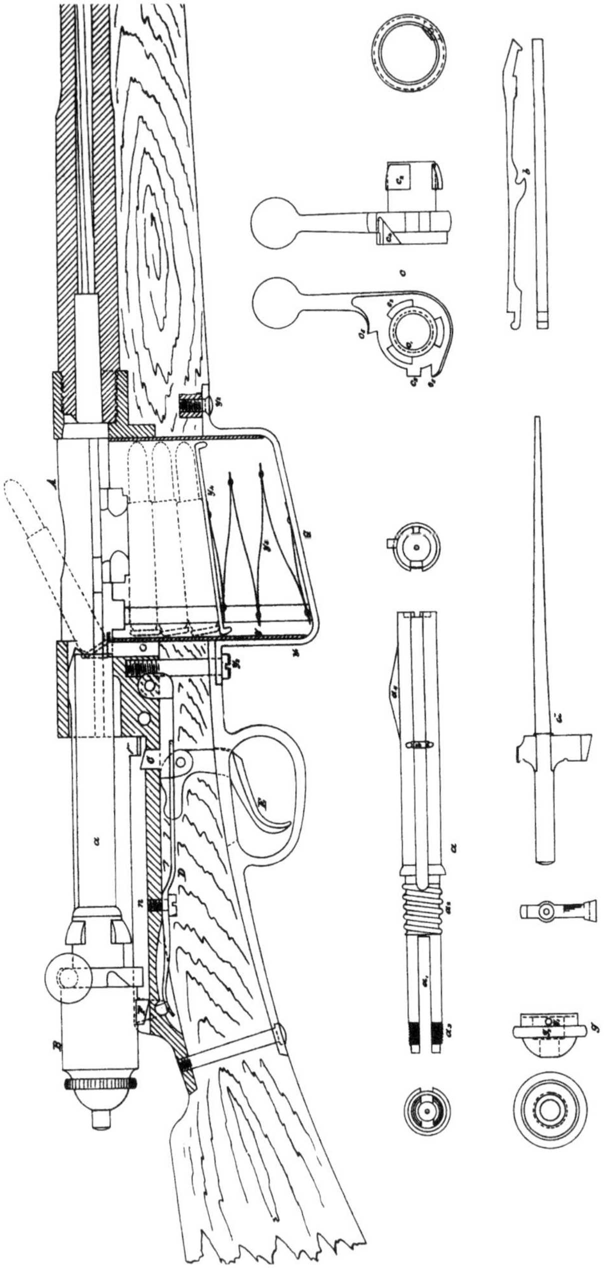

The Rubin rifle was designed by Lieutenant Colonel Eduard Rubin and was entered into the 1888 U.S. Magazine Repeating Rifle trials. The rifle was an improved Vetterli design with a box magazine chambered for a 7.5 mm cartridge very similar to that adopted two years later for the Model 1889 Schmidt rifle.

Zielfernrohr-Karabiner 31/55 (Model 1955 Sniper Rifle) with Kern 3.5 power telescopic sight and carrying case. The rifle has an attached bipod. Ian Shein collection.

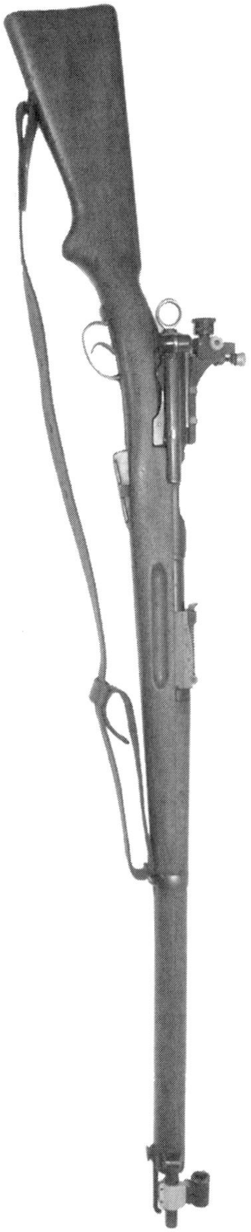

Model 1911 rifle with Anschutz sight package for use in standard Swiss shooting matches. The owner reports that the rifle can drive tacks. Ian Shein collection.

Lt. Colonel Rudolf Schmidt, the developer of the Schmidt rifle system.

Lt. Colonel Eduard Rubin, the developer of the first successful small-caliber, jacketed cartridge for repeating rifles.

Chapter 1

The Advent of the Bolt-Action, Repeating Rifle

Switzerland is a small nation surrounded by high mountains in the cockpit of Europe. The entire nation isn't as large as the combined states of Vermont and New Hampshire. Since the end of the Napoleonic Wars in 1815, it has only fought one civil war in which 128 people were killed, while remaining as neutral as possible through a dozen European wars, including two World Wars. It has done so because of a well-equipped and trained citizen army supplied with excellent weapons.

A Short History of Switzerland

Switzerland became a confederation in 1291 and gained its independence from the Holy Roman Empire and its ruling Hapsburg family in 1499. At the conclusion of the Thirty Years' War in Central Europe, in which the Swiss managed to remain uninvolved, the country was recognized as both independent and neutral by the Treaty of Westphalia in 1648.

In 1798, Napoleon invaded Switzerland and established the Helvetic Republic, which quickly collapsed following the Battle of Waterloo in 1815. The Congress of Vienna, called to reorder Europe after the fighting ended, guaranteed Switzerland's independence and neutrality.

A new democratic constitution in 1848 established the Canton system under a federal government. At the same time, the Swiss people vowed never again to be invaded by a foreign enemy and instituted universal conscription for military service.

Today, all men on reaching their 19th birthday (the conscription age has varied between 19 and 20 since the 1850s) are obliged to spend 15 weeks on active duty in basic training. Some areas of speciality require longer training. After their initial period of active duty, they enter the ready reserve during which they are required to serve specified active duty periods every two years. Depending on rank, reservists are released from further service at ages 42, 52 and 62. Military service for women is voluntary.

The Swiss have developed a very efficient military service in which most of the male population are trained soldiers. Members of the reserve forces are required by law to maintain proficiency with their individual rifles, and most participate in periodic rifle matches. As each soldier is a member of his local military reserve unit and maintains his individual weapon and ammunition ration in his home, the entire armed forces of the Swiss nation can be mobilized within 72 hours.

Small Arms Lineage

On January 8, 1869, Switzerland became the first nation to adopt a magazine, repeating rifle using a self-contained metallic 10.4 x 38 mm (.41-caliber) rimfire cartridge for its armed services, see Figure 1-1. The Model 1869 Vetterli predated the adoption of the Fruwirth carbine for police forces in the Austro-Hungarian Empire by three and one-half years. Professor Friedrich Vetterli of the Schweizerische Industrie-Gesellschaft (SIG) in Neuhausen, Switzerland, had designed and developed a rifle that combined the locking turn-bolt system based on the German Dreyse needlefire rifle with the tubular magazine and elevator assembly of the American Henry and Winchester Model 1866 repeating rifles, see Figures 1-2 and 1-3. The Vetterli magazine rifles replaced the American Peabody single-shot

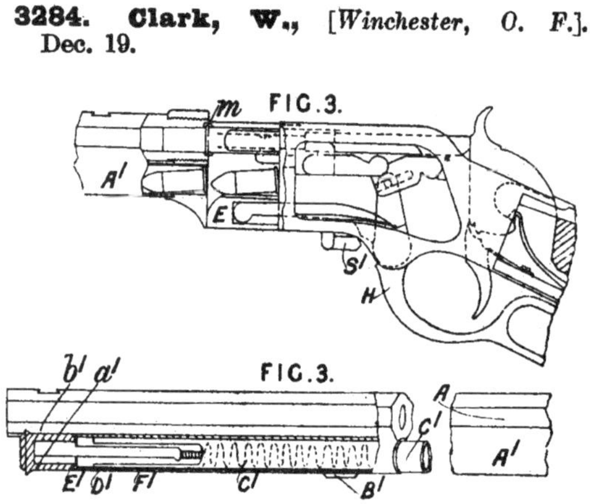

Fig. 1-3. These patent drawings illustrate the magazine and elevator lift mechanisms of the Winchester Model 1866 Rifle. They were adopted into the Vetterli system.

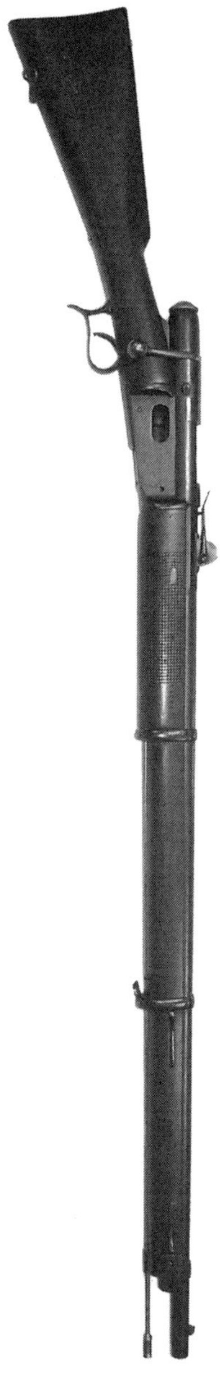

Fig. 1-1. Model 1869 Vetterli Repeating Rifle for Infantry. The loading gate cover and elevator assist mechanism were removed when the rifle was updated. North Cape Publications collection.

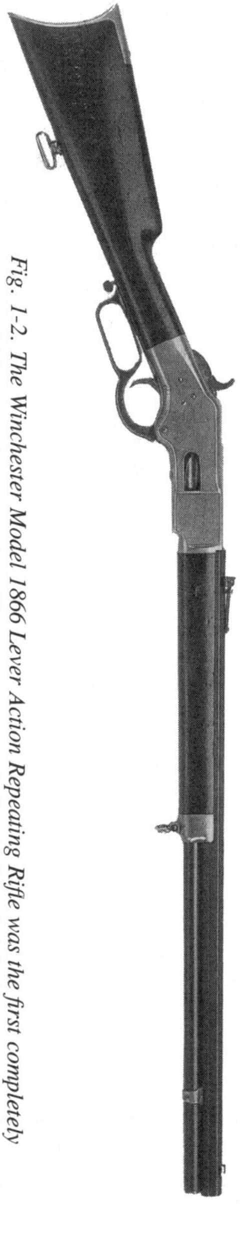

Fig. 1-2. The Winchester Model 1866 Lever Action Repeating Rifle was the first completely successful lever action magazine rifle to be developed. North Cape Publications collection.

metallic cartridge rifles, which had been adopted in 1867 as a stopgap measure until the Vetterli rifles could be produced in sufficient quantity. They also superseded a total of 133,000 percussion muskets that had been converted to fire the same cartridge using the Milbank-Amsler trap door system. Interestingly enough, at the time the 10.4 mm cartridge was considered to be a small-caliber cartridge.

Dreyse's needle gun in the hands of Prussian troops in the 1860s had terrorized Europe. The Prussian juggernaut overran independent Schleswig-Holstein and Austria, and the Germans were eyeing other small nations around them. And it would soon help defeat the French Army in the Franco-Prussian War. The Prussian needle gun was the epitome of infantry small arms in the 1860s, see Figures 1-4 and 1-5.

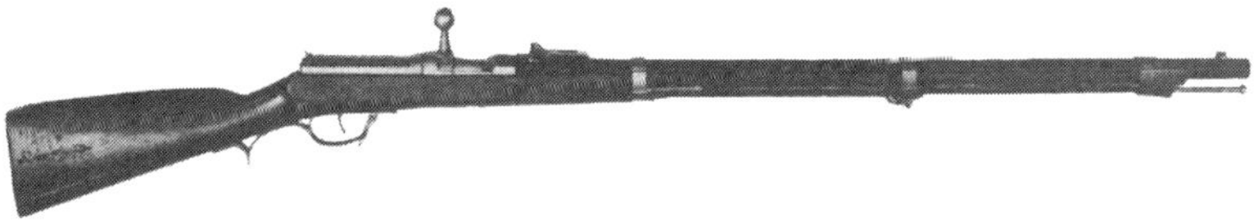

Fig. 1-4. Prussian Dreyse Needle Gun, Model 1868, .60-caliber rifle with a 35-inch barrel. The bolt action was designed for sealing the paper cartridge in the breech. Paul Scarlata collection.

The Dreyse system was the antecedent of the modern bolt-action system. To use, the soldier raised the bolt handle and drew back the bolt to open the breech. This action also pulled the main spring wrapped around the long, needlelike firing pin back and held it in the cocked position with a sear. The soldier then inserted a prepared paper cartridge containing the bullet and powder charge. The priming compound was in the base of the bullet and when he pushed the bolt forward and turned it to the right, the bolt head sealed the breech. When he pulled the trigger, the firing pin was released by the sear and driven forward by the compressed main spring. Its tip penetrated the paper cartridge and drove through the powder charge. It then struck and ignited the

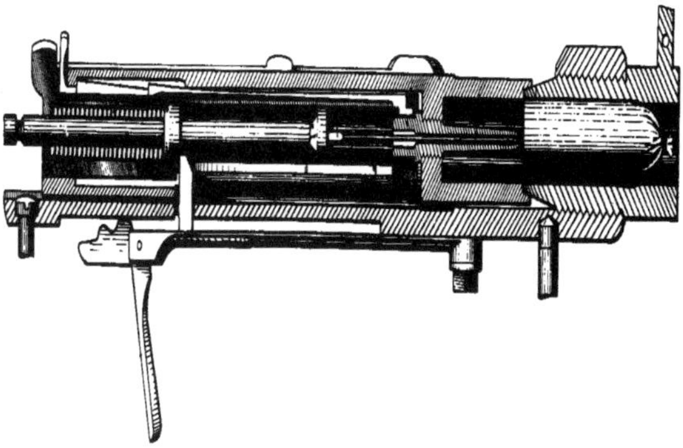

Fig. 1-5. Patent drawing showing how the needlelike firing pin penetrated the paper cartridge to strike the priming compound in the base of the bullet.

priming compound in the bullet's base, which ignited the main powder charge and propelled the conical bullet down the barrel. Reloading was accomplished by opening the bolt and inserting a new cartridge. The rifle was quick to reload and fire and quite reliable for the era of paper cartridges.

When it came to defense, the Swiss had a major geographic advantage that other European nations lacked. The nation was surrounded on three sides by mountain ranges but to the northwest, the land sloped down to a border with Germany, blocked only partially by Lake Constance. It was this plain, the industrial heartland, that was most vulnerable to invasion. The surrounding mountain ranges not only provided a strong defensive barrier, but walled the country off from the rest of Europe and made it difficult, if not impossible, to receive material aid from other countries in time of war. The Swiss government therefore was well aware that their arms industry had to be self-supporting in every way; and their small army would have to field greater firepower than any potential enemy to hold the passes in the south while defending the heartland. Studies conducted by the Swiss army strongly suggested that the best defense was a continuation of the citizen army but reequipped with a magazine repeating rifle.

Magazine Systems

In the 1860s, only one practical magazine system had yet been devised, the straight tube with a spring-loaded follower to push the cartridges into the receiver. The system had been developed in the United States in the 1850s. It had been used in the semi-successful series of Volcanic arms. The system was then much improved and used in both the Henry (1860) and Winchester Model of 1866 rifle and carbine. A workmanlike variation of the magazine tube was used in the Spencer rifle and carbine with a great deal of success. The Henry, Spencer and Winchester Model 1866 all made fine hunting rifles but all had the same three serious drawbacks that limited their military utility. Troops firing from a prone position found it difficult to work the lever action without rolling to the side and possibly exposing themselves to enemy fire. Secondly, the short receivers of both rifles limited the length, and therefore the power, of the cartridge that could be used. Third, the tube magazine with its long spring was vulnerable to malfunctions caused by dirt and grease or if the spring kinked or broke.

Vetterli combined the turnbolt of Dreyse needlefire rifle, the locking lugs of the Greene/Chassepot system but moved to the rear, and magazine and elevator assembly of the Winchester Model 1866 rifle. The problem of dirt in the magazine tube was reduced by enclosing it fully in the forend and by inserting a follower rod to prevent kinking.

Vetterli persuaded the Swiss military authorities to use the 10.4 x 38 mm rimmed cartridge developed for their Peabody Model 1867, and also used in the Milbank-Amsler percussion musket conversions. This small-caliber cartridge would help reduce the weight of the rifle while the rapidity with which the rifle could be fired would overcome the low power of the cartridge. The country would also save the cost of developing a new cartridge. The Vetterli rifle served as a first line issue rifle until 1890. It underwent four model upgrades and was not completely withdrawn from use until the end of World War I.

See Table 1-1 for a comparison of the Vetterli cartridge with others of its time period.

Non-Swiss Historical Notes

The Vetterli was a rugged, serviceable military weapon in spite of its extreme length and weight. The Italian Army adopted a single-shot

Table 1-1 Some European Military Metallic Cartridges, circa 1869, Compared			
Country and Designation	**Bullet Weight**	**Mean Velocity**	**Mean Energy**
Swiss Vetterli 10.4 x 38R mm	334	1427	1330
British Snider .577	350	1310	1338
Egyptian Remington .43	400	1330	1570
Austrian Werndl M73	340	1270	1225
Belgian M1867	386	1368	1610
Russian Berdan 10.75 mm	370	1450	1724
Source: Cartridges of the World, 8th Edition, DBI Books, 1997			

version of the Vetterli rifle in 1870, and later added a box magazine holding four cartridges in 1878.

An interesting connection between Switzerland and Finland occurred during the Russian Revolution of 1905. Finland had been a Russian Grand Duchy since its capture from Sweden in 1809 during the Napoleonic Wars. The Finns had lived in fair harmony with the Swedes for most of their history. The Finns and the Russians merged like oil and water but conciliatory gestures on the part of the Imperial Russian government, plus the granting of a fair degree of independence, prevented the establishment of a serious movement for Finnish Independence until 1899. In that year, the increasingly reactionary Russian government decided to "Russify" Finland.

Swiss Magazine Loading Rifles

After the start of the Russo-Japanese War in 1904, the Japanese covertly supported both Russian and Finnish revolutionary groups. The Russian naval defeat at Tsushima in 1905 and the end of the war focused unrest and led to the first Russian Revolution. Demonstrations and riots broke out in Finland in support of Finnish independence.

The Japanese intelligence service purchased 15,560 obsolete Model 1869 Vetterli infantry rifles from French and German arms merchants who had recently acquired them from the Swiss government. The Vetterlis were loaded aboard a cargo ship, the SS *John Grafton*, which sailed through the North Sea and into the Baltic. As the rifles were being unloaded off the coast of Finland at Pietarsaari, the ship was blown aground by strong winds. When it proved impossible to refloat the ship because of extensive damage, rather than let the Russian Navy confiscate the rifles, the ship was scuttled.

But the Russian Navy managed to salvage most of the rifles and ammunition. Rather than destroy them, they were stored in Russian arsenals in Finland. When the Finnish Civil War — an internal conflict between White nationalists and Red communists — began in 1918, the Vetterli rifles, now known as "Grafton" rifles after the sunken steamer, played a part in the fighting.

A few of the "Grafton" rifles were even used against the Russians by rear area support troops during the Winter War (1939–40) and the Continuation War (1941–1944). In the 1950s, those that had survived a revolution and two wars were sold off as surplus once again.

Many Vetterlis were imported from both Switzerland and Finland into the United States early in the 20th century. Most were converted to sporting rifles by shortening the magazine tube and forend, and often the barrel, so that the cartridge capacity was reduced to five rounds.

The Francis Bannerman catalogs for 1903 and 1906 both list the Model 1869/71 Vetterli rifle (with checkered handgrip). In 1903, the price for one rifle and ten cartridges was $8.00, with an extra 100 cartridges at $2.00 additional. By 1906, the price for the rifle had dropped to $3.95. A sporterized version was offered for one dollar more.

The sporting version of the Vetterli proved so popular that American cartridge makers manufactured a ".41 Swiss" rimfire cartridge using smokeless powder. This often necessitated reworking the front sight to match the ballistics of the smokeless cartridge. The author owns one such Vetterli, a Model 1869, in which the front sight was not only lowered but reinstalled backward and welded in place in the dovetail.

Vetterli System Description

Bolt Assembly

The Vetterli system is distinguished by its bolt and magazine feed system. The bolt was contained within a large-diameter tube atop the receiver that somewhat resembled the Winchester Model 1866 receiver with its right-side oval loading port, see Figure 1-6. The Vetterli bolt was not a true revolving bolt system. Only the bolt collar with the two conical-section lugs rotated when the bolt handle was raised or lowered. When the bolt was drawn back, the extractor withdrew a fired cartridge case from the breech, cocked the spring-loaded firing pin, held it back with a sear and released it when the trigger was pulled. See the exploded view of the bolt in Figure 1-7.

Fig. 1-6. Vetterli bolt (cocked) and housing. Note the "Winchester-like" loading port on this Model 1871 receiver.

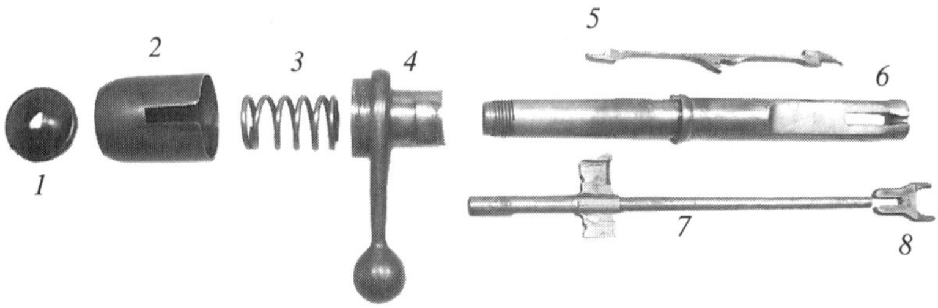

Fig. 1-7. The Vetterli bolt: (1) bolt cap, (2) main spring cover, (3) main spring, (4) bolt handle, (5) extractor, (6) bolt body, (7) striker, (8) rimfire firing pin.

NOTE: The Vetterli's non-rotating bolt body and rotating bolt collar was the inspiration for a number of .22 rimfire rifle systems over the years, the most prominent of which was the Winchester Model 52.

When the lever protruding from the right side was raised, it cocked a spring-driven striker, see Figure 1-8. The striker had a forked firing pin which struck both sides of the cartridge rim of the 10.4 x 38R mm rimfire cartridge to ensure ignition of the priming compound, see Figure 1-9. The extractor was attached to the bolt body and drew the cartridge out of the breech when the bolt was pulled back. The locking lugs were two triangular studs on the rear of the bolt that meshed with similar recesses in receiver, see Figure 1-10.

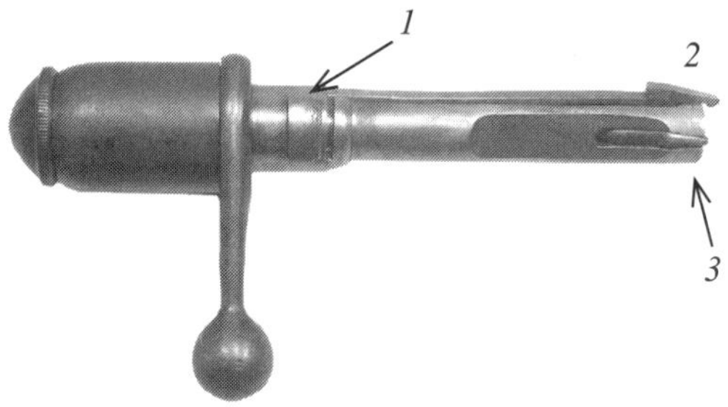

Fig. 1-8. Vetterli bolt showing one of the two (1) locking lugs, (2) extractor and (3) rimfire firing pin.

Magazine and Ammunition Feed System

The tubular magazine was mounted under the barrel, and contained within and protected by the forend, see Figure 1-11. Cartridges were loaded into the magazine by pressing them into a cutout or loading

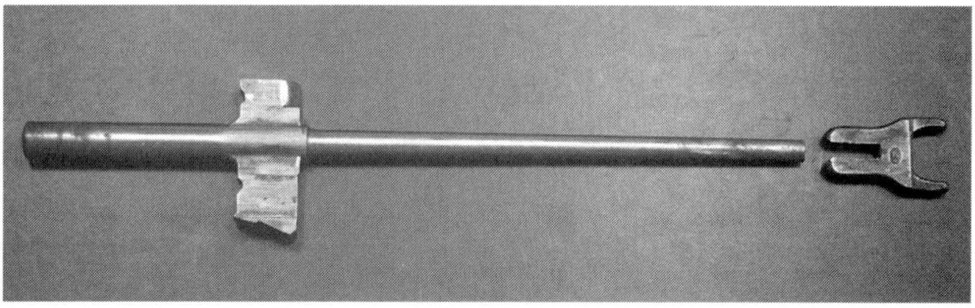

Fig. 1-9. Vetterli striker and twin-prong firing pin for the rimfire cartridge.

port in the right side of the receiver. As each cartridge was inserted, it compressed a coil spring in the magazine tube. The spring was capped by a metal follower which pressed against the nose of the first cartridge, pushing all back toward the receiver. Early Model 1869 Vetterli rifles had a magazine capacity of twelve rounds, but after the first several thousand were manufactured, the capacity was reduced to eleven. Both variations could carry an additional cartridge in the elevator.

Fig. 1-10. Vetterli bolt handle showing the two rear-mounted bolt locking lugs.

When the bolt was pulled back, a groove in the bottom of the bolt cylinder moved back the arm of a bell crank hinged in the receiver, causing the bell crank to rotate and lift an arm on the opposite side which pressed the elevator up with enough force to eject the empty case, see Figure 1-12. The semicircular cut in the rear side of the elevator was shaped to toss the cartridge out to the right rear at a shallow angle. When the magazine was loaded,

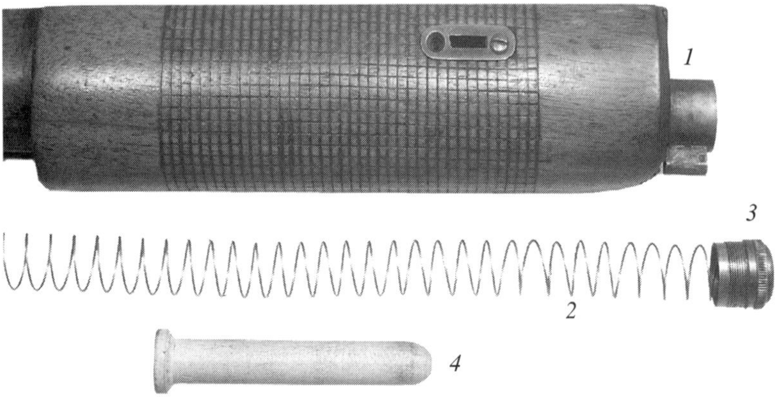

Fig. 1-11. The Winchester-type magazine tube (1) was contained within the forend. A coil spring (2) drove the follower (3) that moved the cartridges into the elevator. The wooden follower plug (4) prevented kinking when the spring was compressed.

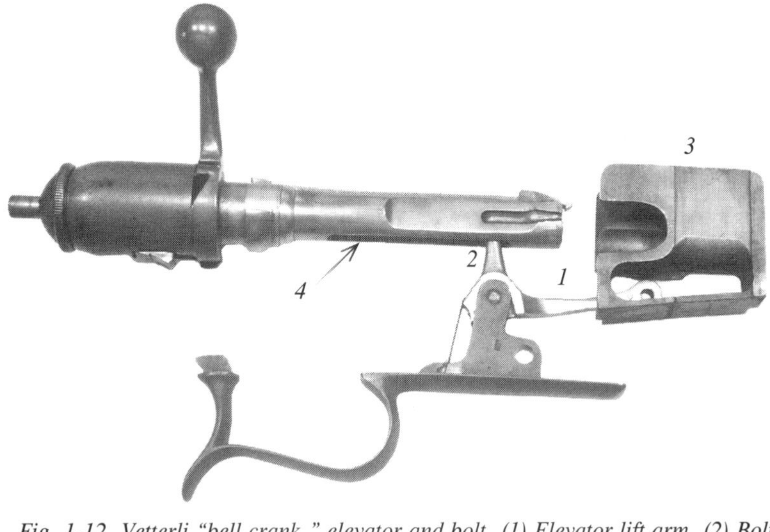

Fig. 1-12. Vetterli "bell crank," elevator and bolt. (1) Elevator lift arm. (2) Bolt arm. (3) Elevator. (4) Groove in the bottom of the bolt body.

the elevator always contained a cartridge due to the pressure of the spring and follower in the magazine.

When the bolt handle was pushed forward, the bolt face drove the cartridge out of the elevator and into the barrel's breech end. At the same time, the movement of the bolt arm in the bolt body

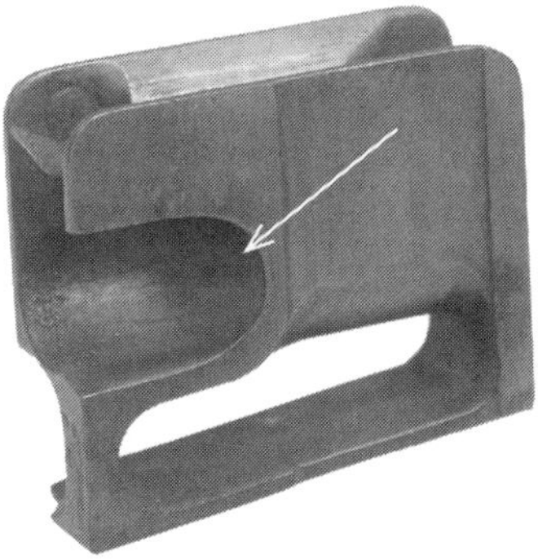

Fig. 1-13. Vetterli "elevator" or lifter block.

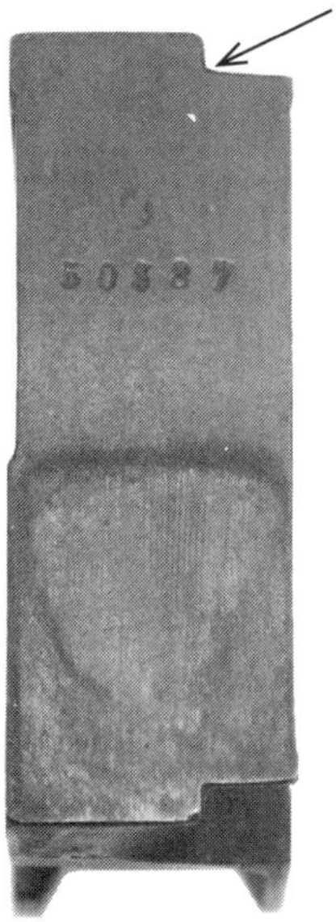

Fig. 1-14. Model 1869 Vetterli elevator. Note the notch for the elevator lift spring that was eliminated in subsequent models.

groove caused the bell crank to rotate in the other direction, drawing the elevator back down to receive a new cartridge.

The elevator was similar in shape to that used in the Model 1866 Winchester (called a carrier block by Winchester), see Figure 1-13. It was rectangular in shape with a "U"-shaped cutout of the right side (arrow) to allow cartridges to pass through when the rifle was being loaded. The bottom of the elevator was open to allow the arm of the bell crank to move it up and down. The inside walls of the elevator were shaped to cradle the 10.4 x 38R mm cartridge, including areas milled away on either side for the cartridge rim. Model 1869 elevators can be identified by a rectangular notch milled on the left, front bottom of the elevator for the elevator lift spring, see Figure 1-14, arrow. This spring was used only in the Model 1869.

Stock, Forend/Forearm and Furniture

The Vetterli had a separate buttstock and forend (rifle) or forearm (carbine). The buttstock was attached to the rear of the receiver; the forend or forearm was slung under the barrel (Figure 1-15) and held with one, two or three barrel bands and a key, depending on the model and type, see Table 1-2.

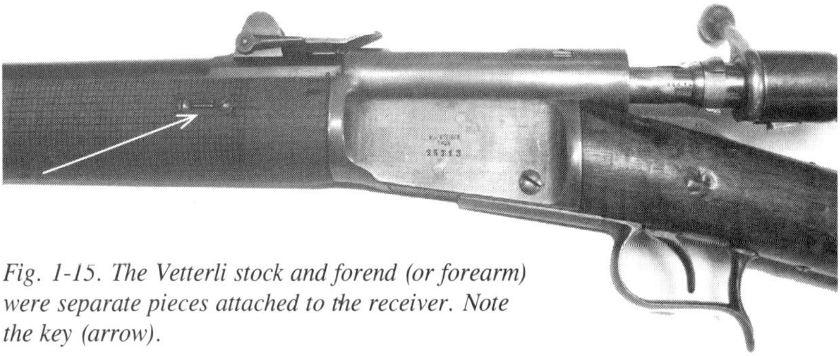

Fig. 1-15. The Vetterli stock and forend (or forearm) were separate pieces attached to the receiver. Note the key (arrow).

The forward band was two inches wide with an opening for the cleaning rod that ran through the center, bottom of the forend or forearm. A Mauser-style bayonet mount was added to the right side in the Model 1878/81 rifles. In the Model 1869/71 a band spring with a locking extension protruded through a hole in the right side of the front band. The rear band was also held by a band spring plus a clamping screw at the bottom. The middle band was held only by a band spring and carried the sling swivel. The forward band spring was eliminated in the Models of 1878 and 1881 and replaced by a cross screw.

The forend or forearm of the Model 1871, 1878 and 1881 Vetterlis was also secured by a key that passed through either side of the forend or forearm and through a notch cut into the bottom of the barrel. The key was removed from left to right, refer to Figure 1-15.

NOTE: A great many Model 1869 forends were later modified with the key to modernize them.

Table 1-2 Forend/Forearm Barrel Bands by Model and Type		
Model	**Type**	**Number of Barrel Bands**
1869	Infantry Rifle	3
1869	Short Rifle	2
1869/1871	Infantry Rifle	3 + Key
1869/1871	Pioneer Carbine	1
1870	Police Carbine	2
1871	Infantry Rifle	3 + Key
1871	Short Rifle	2 + Key
1871	Carbine	2 + Key
1878/1881	Infantry Rifle	2 + Key
1878/1881	Short *(Stutzer)* Rifle	2 + Key
1878	Repeating Carbine	1

The Models of 1878 and 1881 used a new type of lower band. The sides of the band on either side of the split were elongated to form lugs which were drilled through for the lower band screw.

The buttstock lacked a pistol grip but had a pronounced comb that was flat on top. The forend extended to within four inches of the muzzle on rifles and the forearm 1.75 inches on the carbine. A channel for the cleaning rod was cut through the bottom of the forend or forearm. Forends and forearms for the Models of 1869/1871 were checkered in a crosshatch pattern. Forends for the Models of 1878 and 1881 were not checkered.

A transverse wrist pin was passed through the stock wrist starting with the Model 1878 rifle to prevent the stock wrist from splitting when the butt of the rifle was grounded sharply. The transverse pin was made of a hard, dense wood. The ends were notched so that after the pin was placed, wood wedges were driven in to secure it. The transverse pin was also used in the Model 1881 and was retrofitted to many earlier models, see Figure 1-16.

Fig. 1-16. A transverse pin made of wood was used to reinforce the stocks of all Vetterli rifles from the Model 1878 on. Many Model 1869s were also retrofitted with the stock pin.

Two guard screws held the stock to the barreled receiver. One, two or three barrel bands, depending on model and type, held the forend or forearm to the barrel, see Figure 1-17.

In the prototype Model 1868 Vetterli rifle, the cleaning rod channel was cut through the left side of the stock, possibly to avoid complicating the mounting of a bayonet. It was quickly discovered during field tests that the cleaning rods were easily bent and in August 1869, approval was given to mount the cleaning rod in a conventional channel cut along the bottom of the forend or forearm.

The Models of 1869 and 1871 had shallow curved metal butt plates. The Models of 1878 and 1881 had thickened and deeply curved butt plates, refer to Figure 1-17.

Receiver

The Vetterli receiver had a vague resemblance to the Model 1866 Winchester receiver, as might be expected, see Figure 1-18. It was forged from a single piece of steel and machined to final shape. The front of the receiver flared under the barrel mounting to receive the magazine tube. The center of the receiver was cut away for the elevator to move up and down and was open at the top, see Figure 1-19. The top rear of the receiver was shaped into a large tube containing the bolt

Fig. 1-17.

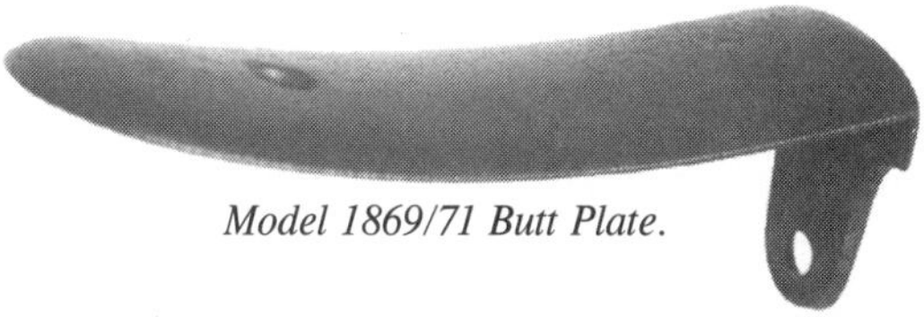

Model 1869/71 Butt Plate.

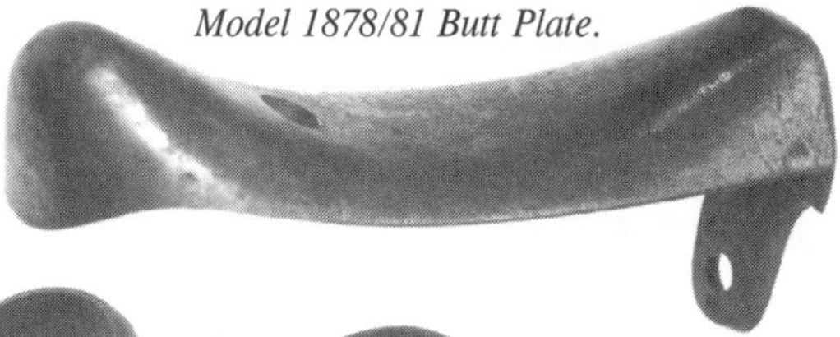

Model 1878/81 Butt Plate.

Model 1878-1881 Front Barrel Band with bayonet mount. The Model 1869/71 Front Barrel Band was the same but without the bayonet mount.

Model 1869/71 Rear Barrel Band.

Model 1869/71 Middle Barrel Band.

Model 1878/81 Rear Barrel Band.

race. Beneath and behind the tube was the top tang for the receiver. The bottom rear of the receiver extended in a tang to the rear and was inletted into the bottom of the stock.

The trigger and sear were mounted in the bottom tang, see Fig-

Fig. 1-18. The Vetterli receiver was cast and forged to shape. Shown above the receiver is an elevator well cover that was fitted to some, but not all, carbines and the Model 1869 rifles to keep dust and water or snow out.

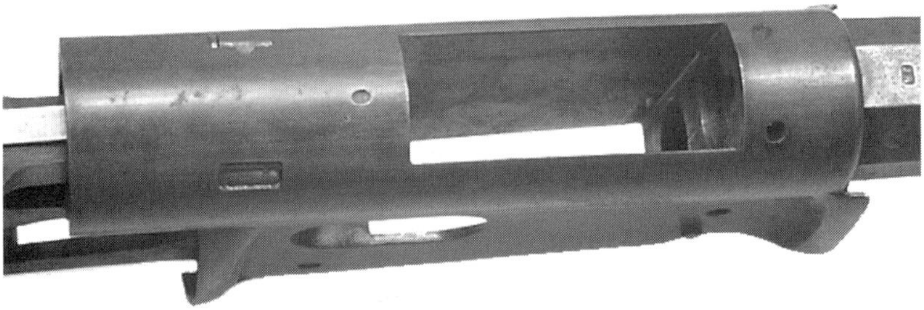

Fig. 1-19. The space above the elevator well was cut away and served as the ejection port for spent cartridges.

ure 1-20. The trigger was pinned in the lower tang, and the sear was pinned to the trigger. When the bolt was moved forward, the sear engaged the cocking lug on the bottom of the striker. When the trigger was pulled to the rear, it drew the sear down out of engagement with the cocking lug on the striker, and the main spring drove the striker and firing pin forward against the base of the cartridge.

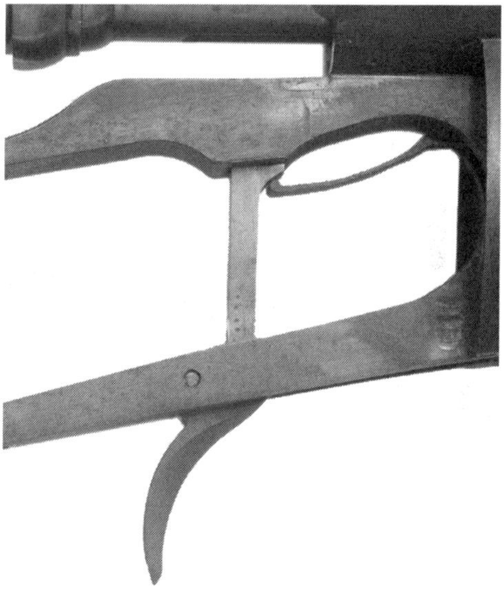

Fig. 1-20. The trigger/sear assembly was pinned into the lower tang.

In the original Model 1869 rifle, a pivoting cover was attached by a screw to the right front of the receiver to close off the loading port and prevent the entry of dust and dirt. The loading port cover was quickly found to be unnecessary and was eliminated in the Model 1871 and later rifles, see Figure 1-21. Two gas ports drilled on either side of the receiver ring were eliminated from the Model 1871 receiver in late 1875.

Also in the original Model 1869 receivers, a spring was attached to the left front inside receiver wall to assist in moving the elevator up. This spring was found to be unnecessary and was eliminated in the Model 1871 and later rifles. The Model 1869 Vetterli also was equipped with a magazine cutoff, a switch that blocked the upward movement of the elevator. The cutoff was eliminated when it was found that the rifle could be single loaded easily by not withdrawing the bolt fully to the rear. The three holes on the left side of the receiver were also eliminated, see Figure 1-22.

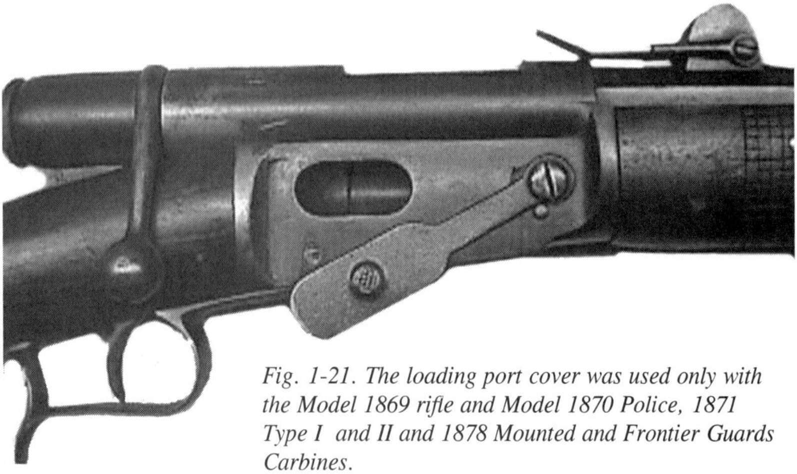

Fig. 1-21. The loading port cover was used only with the Model 1869 rifle and Model 1870 Police, 1871 Type I and II and 1878 Mounted and Frontier Guards Carbines.

Fig. 1-22. Receivers for the Model 1878/81 rifles and carbines lacked the three holes in the left side front for the elevator assist spring and were stamped with the model year.

Most Model 1869s were converted to the Model 1871 receiver configuration while in service. The Model 1869 receiver can be identified by the screw hole and stop for the loading port cover on the right front side of the receiver (refer to Figure 1-21) and by the two indentations and slot on the left front side of the receiver for the elevator assist spring.

At the bottom rear of the receiver, a hole was drilled for the front trigger guard screw which passed through the receiver from left to right, refer to Figure 1-22, arrow.

Trigger Guard and Elevator Lift Assembly

The trigger guard was attached to the bottom of the receiver to protect the trigger and prevent it from being tripped accidentally, see Figure 1-23a. The front part of the trigger guard also contained the elevator lift assembly. The rear of the trigger guard contained a lug (arrow) which was inserted into a slot in the bottom tang to secure it. A cross- bolt passed through both sides of the receiver near the front and through a projection on the lift mount to secure it at the front.

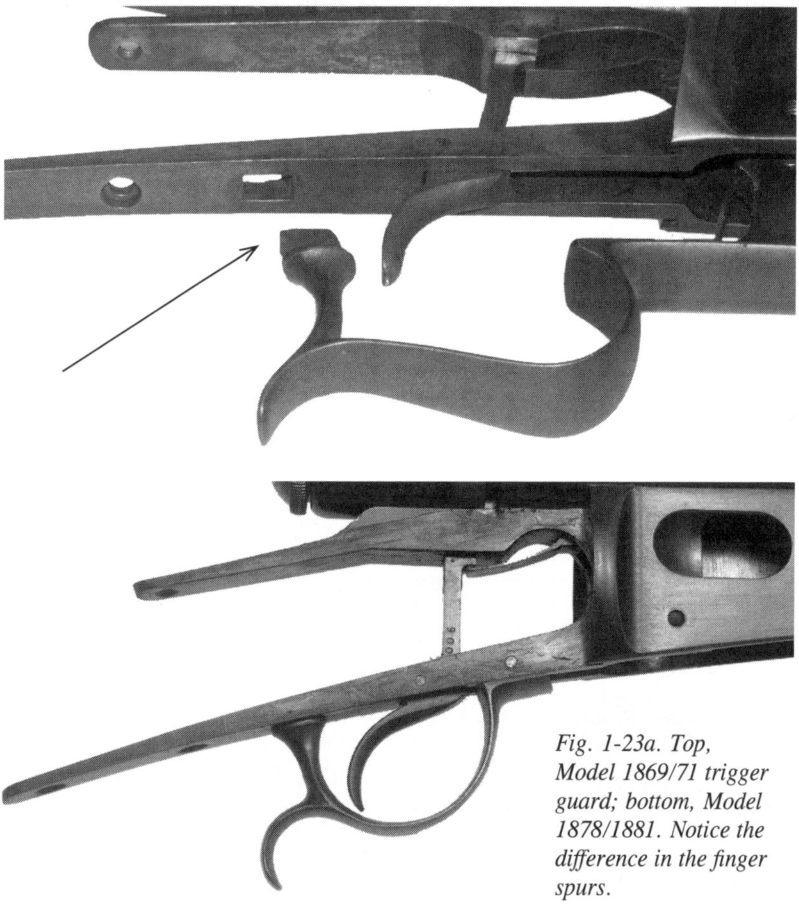

Fig. 1-23a. Top, Model 1869/71 trigger guard; bottom, Model 1878/1881. Notice the difference in the finger spurs.

The trigger guard plate manufactured for the Model 1869/1871 was made in one piece while that for the Model 1878/1881 was made in two pieces. The elevator lift assembly was mounted on a separate plate, see Figure 1-23b, arrow.

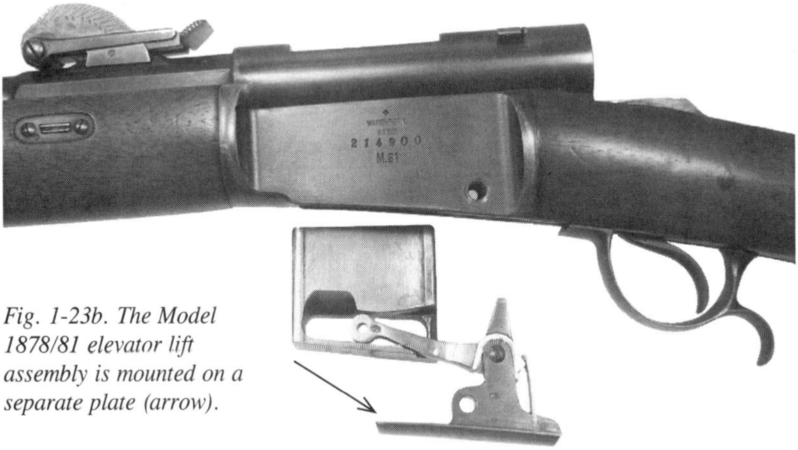

Fig. 1-23b. The Model 1878/81 elevator lift assembly is mounted on a separate plate (arrow).

The elevator lift assembly consisted of the elevator mount, a bell crank which pivoted on the mount, and a leaf spring which assisted its up and down movement, see Figure 1-24. The bell crank lifted the elevator containing a cartridge up and into line with the breech so that the bolt face could push it into the breech, refer also to Figure 1-12.

Fig. 1-24. Close-up of the bell crank assembly that moved the elevator up and down.

The elevator lift assembly manufactured for the Model 1869/1871 shows a great deal of filing and shaping by hand, and the lift arm is 0.150

inch thick at its narrowest point. The Model 1878/1881 elevator lift assembly by contrast is all machine made with little evidence of hand fitting.

The Vetterli rifle trigger guard had a finger extension at the rear. The carbine trigger guard did not have the finger extension. The Model 1869/71 trigger guard was larger than that used on the Model 1878/81 rifle, and the rear of the guard was vertical. The Model 1878/81 trigger guard was rounded with a smaller finger extension, see Figure 1-25.

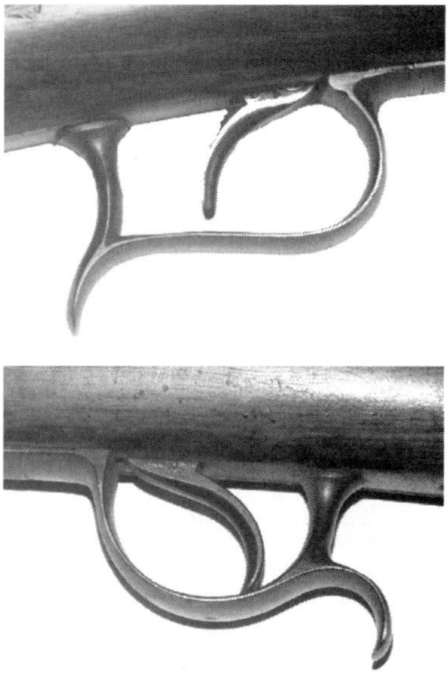

Fig. 1-25. Trigger guards: top, Model 1869/71; bottom, Model 1878/81.

NOTE: Matching Roman numeral fitting numbers have been observed on many bell cranks and elevators. It stands to reason that these parts would have to have been carefully hand fitted to function properly. Elevators or bell cranks lacking fitting numbers may be replacement parts which were fitted at depots or at the unit level one at a time and so did not require an identifying number during reassembly.

Sights

Rear Sight

The rear sight used on the Vetterli rifle and carbine was called a quadrant sight, see Figure 1-26. The base had curved (tangent) rails grooved on the inside of the left side. The leaf was pinned at the front left end of the base and pivoted up and down. To adjust the elevation, the leaf was moved up or down to the chosen elevation. A flat spring was attached to either side of the base and pressed against the base and

Fig. 1-26. Model 1863 rear sight used on the Model 1869 rifle.

side of the leaf to hold it in place at the chosen elevation. The spring pressure was enough to hold the leaf in place against the mild recoil of the 10.4 x 38 mm rimmed cartridge. The rear sight could be adjusted for elevation only and not for windage.

The rear sight used on the Model 1869 rifle was the Model 1863 rear sight, graduated in "Schritt" (paces). It had a short leaf that protruded one inch beyond the base. The left side of the base was graduated from 225 to 1,000 paces.

Fig. 1-27a. The Model 1871 rear sight was a modification of the Model 1863 but with a longer leaf. It was graduated in meters.

The Model 1871 rear sight leaf protruded two inches beyond the base for increased elevation, and the rear sight base was regraduated from paces to meters, see Figure 1-27a. The longer leaf and base graduated to meters was also installed on many Model 1869 rifles sent for refurbishing and repair.

A new tangent-type rear sight was developed for the Model 1878 rifle. It curved forward and around from its dovetail base and was graduated on the left side of the base from 225 to 1,200 meters, see Figure 1-27b. The sight leaf was hinged at the front on a pivot screw. A leaf spring on the left side was secured by the pivot spring and pressed against the side of the base to hold the leaf in position. A stop was milled on the inside of the left side of the base to prevent the leaf from being overextended.

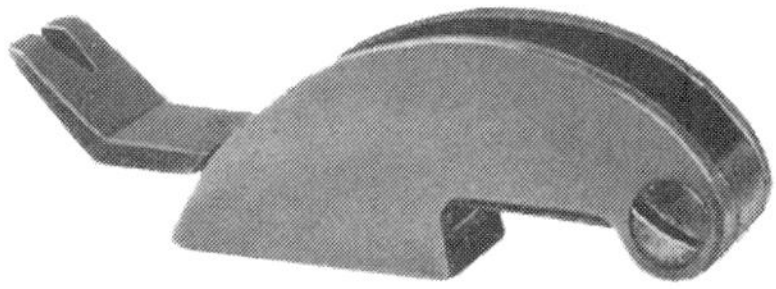

Fig. 1-27b. Model 1878 tangent-type rear sight. It was graduated to 1,200 meters.

The Model 1881 rear sight was quite similar to the Model 1878. The main difference lay in the fact that the aperture was attached to an extension that slid into the leaf. With the sight leaf set at the 1,200 meter mark, the soldier pulled the aperture extension out to expose additional graduations to 1,300, 1,400, 1,500 and 1,600 meters with fifty-meter indices between, refer to Figure 1-28.

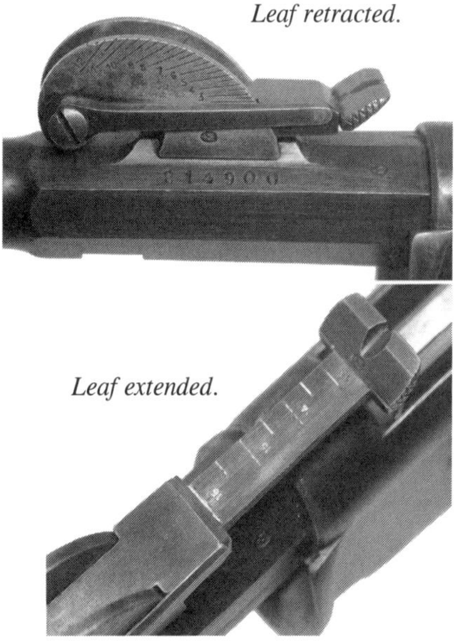

Fig. 1-28. The Model 1881 rear sight was graduated to 1,600 meters. The leaf extended to cover the extra range.

The Model 1869 Vetterli carbine was equipped with a rudimentary fixed sight which had two folding leaves. Most were later replaced with a tangent-style rear sight.

NOTE: The Model 1878 rear sight was replaced with a Model 1881 rear sight on all Model 1878 rifles in service. As a consequence, it is quite rare.

Front Sight

The front sight used on the Vetterli rifle and carbine was a "barley corn" or inverted "V" front sight. The sight blade and base were machined in one piece, see Figure 1-29. The base, in turn, was mounted

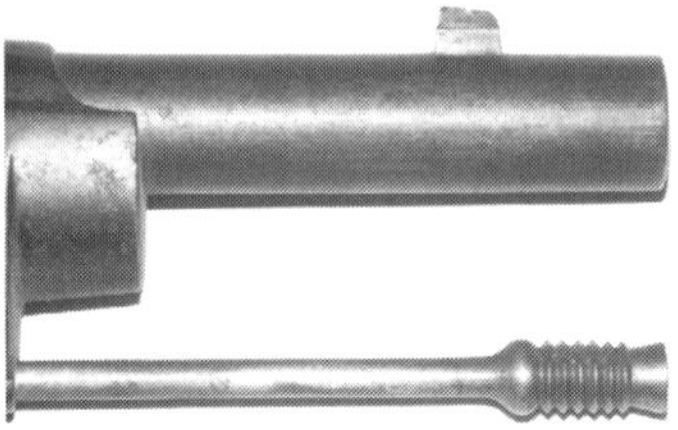

Fig. 1-29. Rifle front sight.

Fig. 1-30. Carbine front sight.

in a dovetail cut into the barrel and brazed in place.

The carbine front sight base and blade was also machined in one piece. But the front sight assembly was mounted on the *front barrel band* rather than the barrel, see Figure 1-30.

Barrel

The Vetterli barrel length depended on the model and type, but all had four grooves and one turn in 26 inches (1:26) with a right-hand twist. Refer to Table 1-3 for barrel lengths and a quick identification of models and types. The barrel was round for its entire length but the rear top six inches ahead of the receiver (the knox form) was ground flat into five faces, see Figure 1-31. The rear sight was inserted into a large dovetail 1.775 inches centered ahead of the receiver face as shown in Figure 1-31.

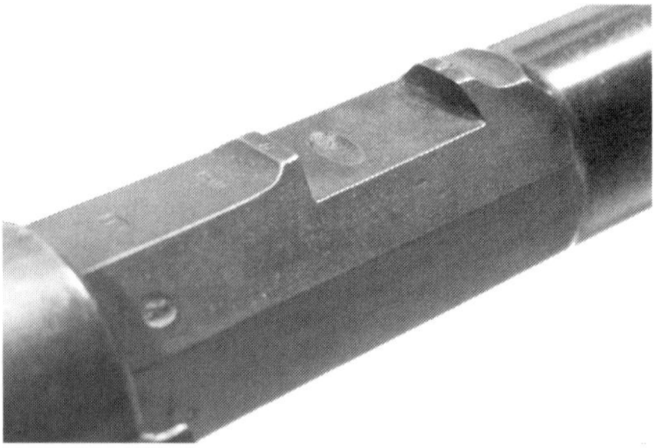

Fig. 1-31. Vetterli barrel showing knox form and dovetail for the rear sight.

The front sight was mounted in a dovetail cut 1.29 inches behind the muzzle. A straight-sided cut was made in the bottom of the barrel 0.465 inch ahead of the receiver face to allow the forend key to pass.

All Vetterli barrels received a deep funnel- or "V"-shaped crown, see Figure 1-32.

As with all major metal parts of the Vetterli, the barrels were nicely polished and blued.

Fig. 1-32. All Vetterli barrels had a deep "V"-shaped crown.

CLEANING ROD

A metal rod was used to both clean the bore and remove stuck cartridges, see Figure 1-33. Referred to as a cleaning rod, as used in the rifle, it was 32.5 inches long. The diameter of the rod was 0.210 inch. It flared into a head, 0.375 inch in diameter. The head was pierced with a slot 0.410 inch long by 0.125 inch wide for a cleaning cloth, see Figure 1-34.

Fig. 1-33. All Vetterli rifles or carbines carried a cleaning rod in the forend or forearm.

In the Prototype Model 1868, the cleaning rod passed through a lug in the upper band and into a channel cut into the left side of the forend. But in this position, the cleaning rod was easily damaged. The cleaning rod channel was moved to the bottom of the forend as was the lug on the upper band, where it remained in all subsequent models and types. The opposite end of the cleaning rod was threaded (Figure 1-35) and screwed into a retaining plate in the stock, just ahead of the receiver.

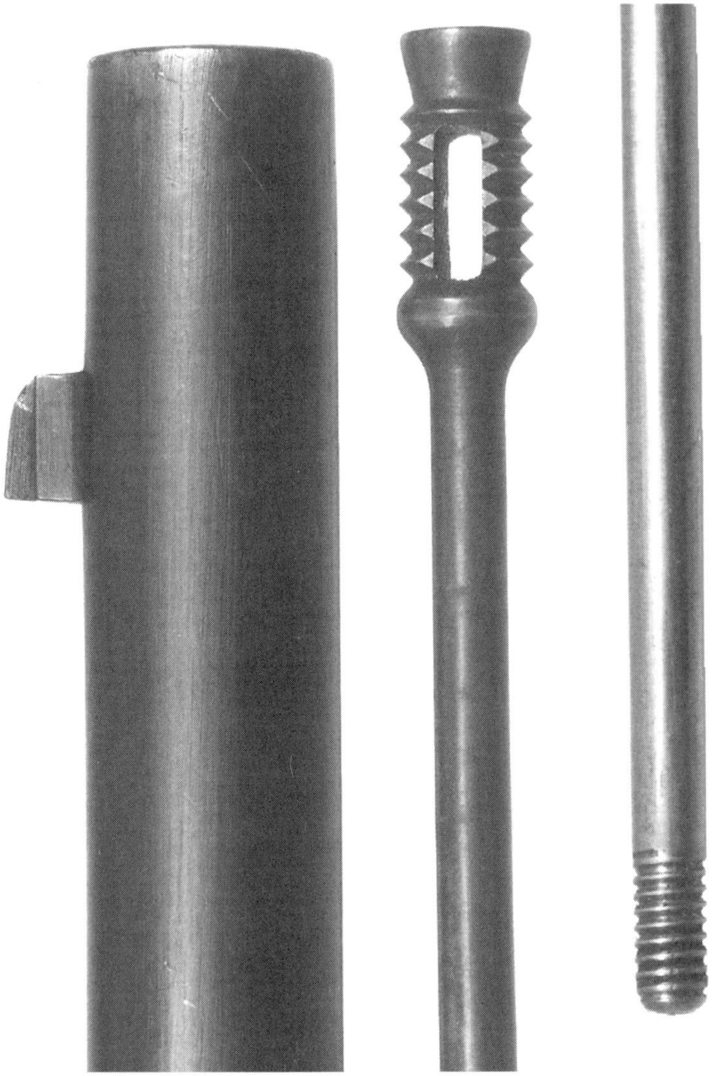

Figs. 1-34 and 1-35. The Vetterli cleaning rod head was slotted for a cleaning cloth, and the end was threaded to screw into the cleaning rod retaining plate embedded in the stock.

Table 1-3 The Issue Swiss Vetterli Rifles, Short Rifles, and Carbines		
Model	**Type**	**Identifying Characteristics**
1869	Infantry Rifle	51.9" long overall, 33.1" barrel, loading gate cover, elevator spring, 1,000-"Schritt" rear sight, shallow curved butt plate, receiver cover, 10.3 lbs, eleven-round magazine, forend crosshatched
1870	Cadet Rifle Type 1	One-piece stock, single shot, 26.75" barrel, OA, 45.25", 7.2 lbs empty, shallow curved butt plate, receiver cover
1870	Cadet Rifle Type 2	One-piece stock, single shot, 26.75" barrel, OA, 45.25", 7.2 lbs empty, shallow curved butt plate, no receiver cover
1870	Police Carbine	45" OA, 29-inch barrel, loading gate cover, no elevator assist spring, reinforcing key in forend, shallow curved butt plate, crosshatched forend, 1863 type rear sight graduated to 800 Schritt (changed to 600 meters), magazine well sliding cover, forward sling swivel on stock (underneath), no bayonet mount, seven cartridges
1869/71	Pioneer Carbine (Konstruktion)	37.2" OA, 19.1-inch barrel, no loading gate cover, no elevator assist spring, reinforcing key in forearm, shallow curved butt plate, crosshatched forearm, 1871 type rear sight graduated to 225 meters, no magazine well sliding cover, forward sling swivel on bottom of rear band, rear sling swivel on buttstock, six cartridges, no bayonet mount
1871	Infantry Rifle	51.2" OA, 33.2" barrel, no loading gate cover or elevator assist spring, reinforcing key in forend, 1,200-meter rear sight, shallow curve butt plate, eleven-round magazine, crosshatched forend, three barrel bands

Table 1-3, cont. The Issue Swiss Vetterli Rifles, Short Rifles, and Carbines		
Model	**Type**	**Identifying Characteristics**
1871	Short Rifle (Stutzer)	30.8" barrel, double-set triggers, no loading gate cover, reinforcing key in forend, deeply curved butt plate, transverse reinforcing pin added to stock wrist, crosshatched forend, ten cartridges
1871	Carbine, Type 1	36.7" OA, 18.5" barrel, 7.2 lbs empty, 6-round tubular magazine, loading gate cover, reinforcing key in forearm, shallow curved butt plate, straight forearm with one barrel band without bayonet mount; early models had leaf rear sight, replaced by tangent sight, spoon bolt handle, transverse reinforcing pin added to stock wrist, elevator assist spring, cavalry sling ring
1871	Carbine, Type 2	36.7" OA, 18.5" barrel, 7.2 lbs empty, 6-round tubular magazine, loading gate cover, reinforcing key in forearm, shallow curved butt plate, stepped and cross-hatched forearm with only the front barrel band used, no bayonet mount, tangent sight, conventional ball bolt handle, transverse reinforcing pin added to stock wrist, cavalry sling ring
1878	Infantry Rifle	52.2" OA, 33.8-inch barrel, no loading gate cover, reinforcing key in forend, 1,200-meter rear sight, deeply curved butt plate, transverse reinforcing pin stock wrist. Cross-hatching on forend omitted, twelve-round magazine tube

Table 1-3, cont. The Issue Swiss Vetterli Rifles, Short Rifles, and Carbines		
Model	**Type**	**Identifying Characteristics**
1878	Short Rifle (Stutzer)	52.2" OA, 32.8-inch barrel, no loading gate cover, reinforcing key in forend, 1,200-meter rear sight, double-set trigger, deeply curved butt plate, transverse reinforcing pin in stock wrist, cross-hatching on forend omitted, twelve cartridges in magazine
1878	Carbine, Mounted	36.6" OA, 18.5-inch barrel, receiver loading port cover, transverse reinforcing key in forearm, tubular magazine holding six cartridges, ball-tipped bolt handle, cavalry sling ring on underside of stock wrist, M1878 quadrant 600-meter rear sight, crosshatched, stepped forearm, forward barrel band with integral front sight, shallow curve butt plate
1878	Frontier Guards Carbine	37.2" OA, 19.1-inch barrel, receiver loading port, reinforcing key in forearm, remade from Model 1871 infantry rifles, shallow curved butt plate, transverse reinforcing pin in stock wrist, five cartridges in magazine
1881	Infantry Rifle	51.9" OA, 33.2-inch barrel, no loading gate cover, 1,600-meter rear sight, reinforcing lug in forend, deeply curved butt plate, transverse reinforcing pin in stock wrist. No cross-hatching on forend. Twelve cartridges in magazine.
1881	Short Rifle (Stutzer)	51.9" OA, 33.2-inch barrel, no loading gate cover, reinforcing key in forend, 1,600-meter rear sight, double-set trigger, deeply curved butt plate, transverse reinforcing pin in stock wrist. No cross-hatching on forend. Twelve cartridges in magazine

Swiss Vetterli Ammunition

Table 1-4
Swiss Vetterli Ammunition[1,2]

Designa-tion	Dimension (mm)	Bullet Type/Weight/Case	Velocity (fps)
GP 1867 [3]	10.4 x 38	lead, round nose, 313 grains. 0.425" diameter x 1.02" long, case length 1.5", rim diameter 0.622", head diameter 0.539", neck diameter 0.444"	1,427[4]
GP 1871	10.4 x 38	lead, round nose, 310 grains, 0.425" diameter	1,427
GP 1878	10.4 x 38	lead-tin, round nose, paper patched, 312 grains, 0.419" diameter	1,427
GP 1880	10.4 x 38	lead-tin, round nose, paper patched (extended to cover case neck), semi-smokeless or possibly early smokeless powder (rauchschwache)	1,427

1. Considered a "small-bore" cartridge at the time.
2. All Vetterli ammunition used a propellant of composition unknown to the author.
3. Abbreviation of "Gewehr Patrone" (rifle cartridge).
4. Chamber pressure for all Vetterli and Peabody ammunition is 20,000 psi.

NOTE: Ammunition styled .41 Swiss Rimfire was manufactured in the United States and some European countries from the early 1900s to the mid-1930s for use in Vetterli rifles and carbines that had been sold as surplus for hunting arms. This type of ammunition was loaded with various types of semi-smokeless and smokeless powders and was not manufactured for or under the supervision of the Swiss government.

Unfortunately, the author could not find an ammunition company anywhere in the world that still manufactures the .41 Swiss.

Model 1869 Vetterli, Infantry Rifle

The Model 1869 Vetterli infantry rifle weighed 10.3 lbs empty and was 51.9 inches long overall. It had a 33.1-inch-long barrel. Five flats were ground onto the sides and top over the chamber (referred to as the "knox form"). The barrel was round beneath the flats and continued round in a straight taper to the muzzle, see Figure 1-36. The infantry rifle had three barrel bands with the sling swivel attached to the bottom of the middle band.

The Model 1868 prototype rifle held twelve cartridges in the magazine, plus one in the chamber. The load was reduced to eleven cartridges in the magazine, plus one in the chamber in the Model 1869 and 1871 rifles.

The Model 1869 and 1871 rifles had a buttstock with a shallow curved metal butt plate, a pronounced comb and a straight wrist. The forend extended to within four inches of the muzzle and was checkered in a square-cut pattern between the receiver and the rear barrel band. The checkering is somewhat uneven and was obviously done by hand. The rear sight used a flat base with curved sides. A leaf with a "V"-notch aperture pivoted on a screw at the front of the base and rotated up the curve to the desired elevation. The leaf was held in place at the selected elevation by leaf springs on either side that pressed against the base and leaf. The base was graduated from 225 to 1,000 "Schritt" (paces) and so marked on the left side. The front sight was the "barley corn"-type dovetailed into the barrel and brazed. The trigger guard had a finger extension on the bottom rear.

Fig. 1-36. Model 1869 Vetterli Infantry Rifle.

The significant identifying feature of the Model 1869 Vetterli infantry rifle was the loading gate cover on the right side of the receiver. It was spring-loaded, paddle-shaped, inletted into the receiver's right outer wall and secured by a large screw. The receiver was marked on the left side with the name of the manufacturer and serial number. The model year was not marked, see Figure 1-37.

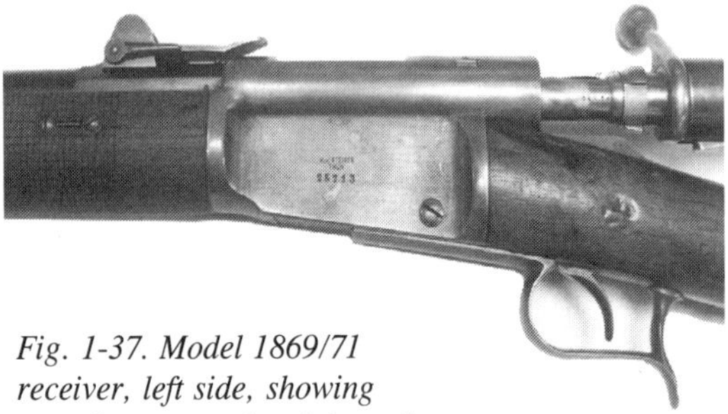

Fig. 1-37. Model 1869/71 receiver, left side, showing manufacturer and serial number.

Prototype Model 1868 infantry rifles had cleaning rods on the left side of the stock. But it was quickly discovered that the cleaning rod was easily damaged in that position and it was then mounted under the stock in the Model 1869 and all subsequent models.

The ring around the bolt cap was originally smooth. After the first several thousand rifles were manufactured, the ring was knurled to make it easier to grasp while unscrewing the cap. Two gas escape holes were also drilled just forward of the opening for the elevator at the two and eleven o'clock positions in the Model 1869.

The Model 1863 angular bayonet, which slipped over the muzzle and locked against the front sight, was issued with all Model 1869/71 rifles. An estimated total of 12,050 infantry rifles were manufactured at eight different factories, as shown in Table 1-5.

Model 1869 Vetterli, Short Rifle (*Stutzer*)

There was no official Short Rifle model designated the Model 1869. An experimental rifle, the Model 1869 *Stutzer* (short), had a 33.2-inch barrel and the Model 1869/1870 had a 30.8-inch barrel.

Both had deeply curved butt plates and were made in very small quantities, probably for evaluation only.

Table 1-5 Model 1869 Vetterli Rifle Factories		
Factory	**Production Years**	**Rifles Manufactured**
SIG, Neuhausen	1869–1874	59,000
Eidgenössische Montier-Werkstätte, Bern	1869–1875	8,900
Eidgenössische Waffenfabrik, Bern	1875–1879	14,060
Cordier & Cie, Bellefontaine	1869–1873	4,000
W. von Steiger, Thun	1869–1874	15,200
Ost-Schweizerische Buschacher, St. Gallen	1869–1874	8,700
Rychner & Keller, Aarau	1869–1873	8,700
Valentin Sauerbrei, Basel	1869–1873	7,000
Zürich Zeughaus, Zürich	1869–1874	1,500

Model 1870 Vetterli, Cadet Rifle

The cadet rifle was a single-shot, smaller, lighter rifle used by school boys in preparatory military training. The rifles used the Vetterli bolt system but lacked the magazine tube and elevator system of the repeating rifle. Their most distinctive feature was the one-piece stock with the receiver contained entirely within the wood. The Cadet rifle did not have a bayonet mount. The exact number manufactured is not known but they were produced by SIG, Neuhausen between 1870–1873, see Figure 1-38.

Two variations were manufactured. The Type 1 had a revolving receiver cover with a knob on the left side. The cover could be moved to the right or left to cover or uncover the magazine well opening.

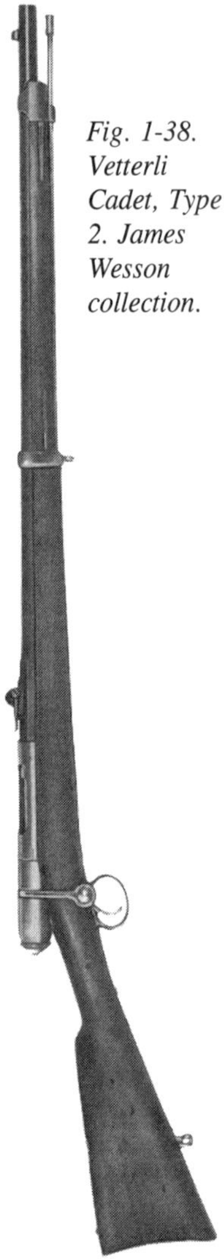

Fig. 1-38. Vetterli Cadet, Type 2. James Wesson collection.

The Type 2 did not have the cover. The Model 1870 Cadet remained in service until replaced in 1897 by the Schmidt Cadet rifle.

Model 1871 Vetterli, Infantry Rifle

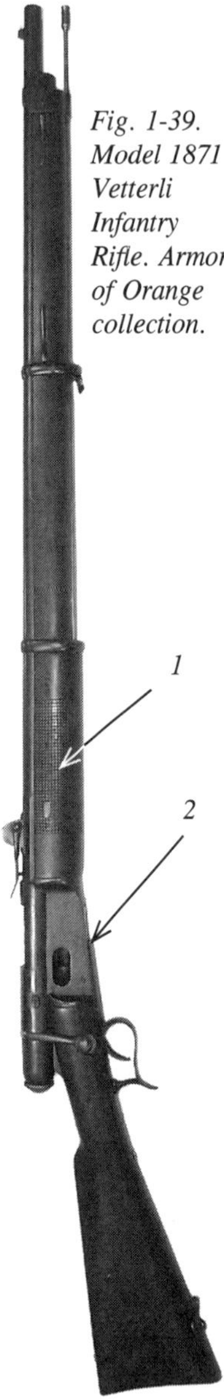

Fig. 1-39. Model 1871 Vetterli Infantry Rifle. Armory of Orange collection.

Several changes were made to the Model 1871 Vetterli, see Figure 1-39. The most visible were the addition of a key (arrow 1) crosswise through the forend below the knox form and the elimination of the loading port cover as well as its screw hole and stop (arrow 2). The elevator lift spring was also eliminated along with the slot and two holes on the left front side of the receiver.

A transverse pin was threaded crosswise through the stock wrist on the Model 1871 and all subsequent models and variations to prevent splitting when the rifle butt was grounded sharply.

The Model 1871 was slightly shorter at 51.2 inches overall with a 33.2-inch barrel. Another noticeable change was the absence of the loading port cover which had been eliminated as unnecessary in the rifle. The elevator lift spring and magazine cutoff was also eliminated. The shallow, curved butt plate as used in the Model 1869 was retained. The rear sight leaf was regraduated from 1,000 "Schritt" to 1,200 meters. The Model 1871 angular bayonet, which slipped over the muzzle, was issued with this rifle.

NOTE: The transverse stock bolt was retrofitted to many Model 1869 rifles, short rifles and carbines.

Like the Model 1869, the left side of the receiver was stamped with the name and address of the manufacturer and the serial number. The model year was not marked, refer to Figure 1-37. The two gas escape holes on the receiver ring were discontinued in 1875.

Model 1871 Vetterli, Short Rifle (*Stutzer*)

The Model 1871 Vetterli short rifle was designed for use by special "Jaeger" battalions, see Figure 1-40. These were light rifle divisions trained primarily for combat in the mountains surrounding the nation. Swiss Jaeger troops were specially trained in marksmanship, and the Model 1871 short rifle used a double-set trigger for more accurate shooting (arrow 1). The *Stutzer* was similar to the infantry rifle but had a barrel slightly shorter at only 30.8 inches. Only two barrel bands were used on the short rifle, the forward band and the middle band. The forward barrel band did not have the "Mauser-style" bayonet stud on the right side. The Model 1871 angular bayonet was used instead. A deeply curved, thick butt plate, of the type used

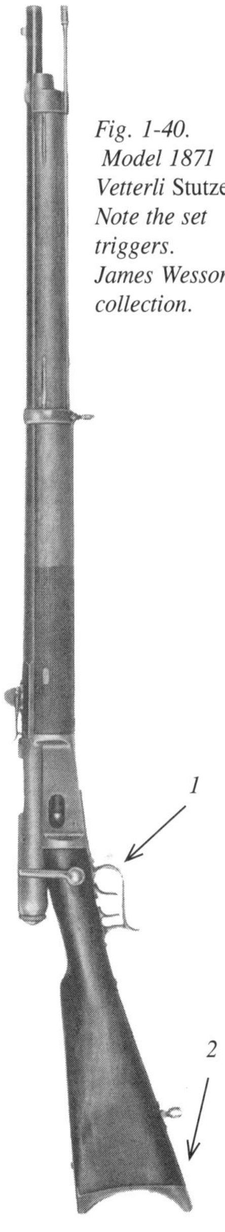

Fig. 1-40. Model 1871 Vetterli Stutzer. *Note the set triggers. James Wesson collection.*

on the experimental development models preceding the Model 1869, was installed (arrow 2). As with the infantry rifle, the Model 1871 short rifle had the reinforcing key in the forend and the transverse pin through the wrist, no loading port cover and no elevator lift spring. A total of 16,000 Model 1871 Short Rifles were manufactured by two factories: Eidgenössische Montier-Werkstätte, Bern and Eidgenössische Waffenfabrik, Bern between 1872 and 1879 (15,000) and SIG, Neuhausen from 1871 to 1874 (1,000).

Model 1870/1871 Vetterli Carbines

Three kinds of Model 1870/1871 carbines were developed and issued, for police, construction or pioneer troops, and for mounted troops.

Model 1870 Vetterli Carbine, Police - The Police carbine was 45 inches long overall with a 29-inch barrel. It had a rifle-type forend, see Figure 1-41. It was fastened to the rifle with a key, the Model 1870 rear clamping barrel band with sling swivel and the forward rifle barrel band but without the Mauser-style stud bayonet mount on the left side.

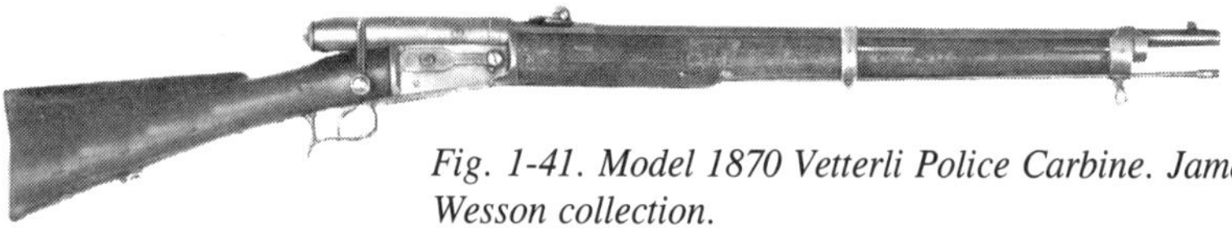

Fig. 1-41. Model 1870 Vetterli Police Carbine. James Wesson collection.

Instead, the police carbine used the Model 1863/67 angular bayonet. The front sight was mounted on the barrel. The rear of the forend was crosshatched. It retained the loading gate cover but not the elevator assist spring. A cleaning rod, reduced in size to fit, was housed under the barrel in the conventional manner. The rear sling swivel was mounted on the bottom of the buttstock and the shallow curve butt plate was used. The police carbine magazine held seven cartridges.

Model 1869/1871 Vetterli Carbine, Konstruktion - The carbine developed for "*Konstruktion*" (pioneer) troops was shorter than the police carbine at 37.2 inches, see Figure 1-42. The barrel was 19.1

inches long and the magazine tube held six cartridges. The front sight was machined from the top of the front barrel band. The loading gate and elevator assist mechanism were both eliminated and a bayonet was not issued. In all other respects, it was similar to the police carbine.

Fig. 1-42. Vetterli Carbine for Konstruktion *Troops. Armory of Orange collection.*

Model 1871 Vetterli, Repeating Carbine, Types 1 and 2

This carbine was developed especially for mounted troops of the cavalry and artillery. It was 36.7 inches long overall and had an 18.5-inch barrel. The Type 1 had a sliding cover over the magazine well. The bolt handle was shaped like that found on a sporting rifle and characterized as the "butter knife" style. The forearm was straight from receiver to muzzle and held with a single barrel band at the front and a cross key at the rear. The front sight was machined from the top of the forward barrel band. A tangent sight rear sight replaced the leaf rear sight found on early Type 1 carbines. A transverse reinforcing pin was added to the stock wrist. A large ring was attached to the bottom of the stock wrist for the cavalry sling.

In May 1872, the Type 2 was introduced. It had a new rear tangent sight (refitted to the Type 1 as well), and the bolt handle was changed to the conventional grasping knob as used on the rifle. The forearm was redesigned with a step but a second barrel band was not used. The forward barrel band was similar to that used on the rifle but slightly smaller in diameter. All other changes made to the infantry rifle and short rifle were made to the carbine as well including the addition of the transverse pin to the stock. The loading port cover was retained, however, but the elevator lift mechanism, slot and two holes on the left front side of the receiver were eliminated, see Figure 1-43.

Some 3,000 Model 1871 carbines of both types were manufactured by two companies in the early 1870s. It is believed that the Type 1 was made at the Eidgenössische Montier-Werkstätte in Bern in 1871 (500) and Type 2 starting in 1872 by Rudolph Pfenninger at Stäfa from 1872 to 1874 (2,500).

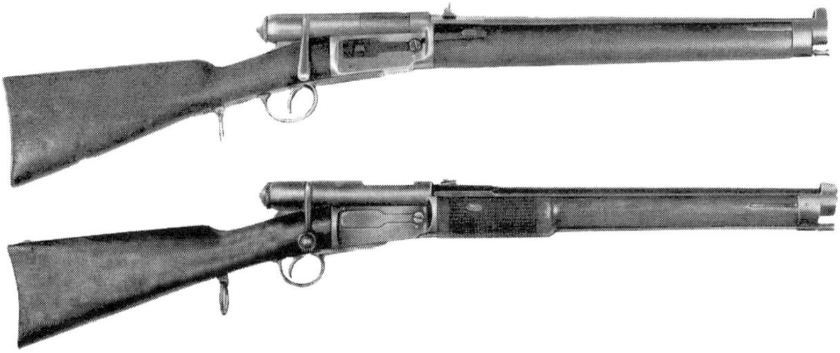

Fig. 1-43. Model 1871 Vetterli Carbines. Above, Type 1; below, Type 2. Note the difference in bolt handles and forearms. James Wesson collection.

Model 1878 Vetterli, Infantry Rifle

The new model Vetterli rifle adopted on April 30, 1878, was really a modified 1871 design. The basic system remained unchanged, see Figure 1-44. The stock was changed to eliminate both the rear barrel band and the cross-hatching on the forend and to provide for the more deeply curved butt plate first used on the Model 1871 *Stutzer* rifle. An improved tangent rear sight was developed that made it easier for the soldier to select the elevation. Unfortunately, nothing was done to improve the shallow "U" sighting notch. The rear sight base was graduated from 225 to 1,200 meters.

The rifle was 52.2 inches long overall with a 33.8-inch barrel. Neither the loading gate cover nor the early elevator lift assist mechanism were used. Magazine capacity was increased to twelve cartridges.

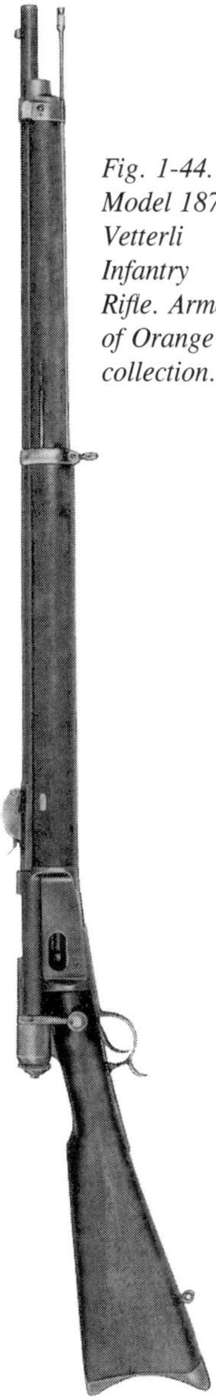

Fig. 1-44. Model 1878 Vetterli Infantry Rifle. Armory of Orange collection.

The forward barrel band was modified with the addition of a "Mauser-style" bayonet stud on the right side. The bayonet issued was the Model 1878 Saber Bayonet with saw teeth on top rear of the blade (see Chapter 6).

The receiver on the Model 1878 Vetterli was stamped with the manufacturer's name and address, serial number and the *model year*, see Figure 1-45.

Fig. 1-45. Left side of the Model 1878/81 receiver was marked with the manufacturer, serial number and model year.

Model 1878 Vetterli, Short Rifle (*Stutzer*)

The Model 1878 Short Rifle retained all the features of the Model 1878 infantry rifle including the deeply curved butt plate designed for the Model 1871 *Stutzer* rifle and the 1,200-meter rear sight. It was 52.2 inches long overall with a barrel 32.8 inches long. The Model 1878 Saber Bayonet with saw teeth was also issued with this rifle. Magazine capacity was increased to twelve.

Some 100,000 Model 1878 rifles were built between 1878 and 1881 at the Eidgenössische Waffenfabrik, Bern.

Model 1878 Vetterli Carbines

Several types of carbines were manufactured for mounted troops, police and frontier guards. Those for military forces were newly manufactured and issued to Dragoon regiments, while those issued to the police and frontier guards were largely remade from the Model 1869/71 infantry rifle with new rear sights, the reinforcing key added to the forearm, an added barrel band with the forward sling swivel and a rifle-type nose cap with a bayonet lug. The Model 1871 butt plate with the shallow curve was retained. It is not known to the author how many Model 1871 rifles were so converted.

The Model 1878 carbine was adopted on April 30, 1878, see Figure 1-46. In addition to the model year on the left side of the receiver on newly built carbines, the loading gate cover was retained but the spur on the trigger guard was eliminated. The nose cap did not carry a bayonet lug.

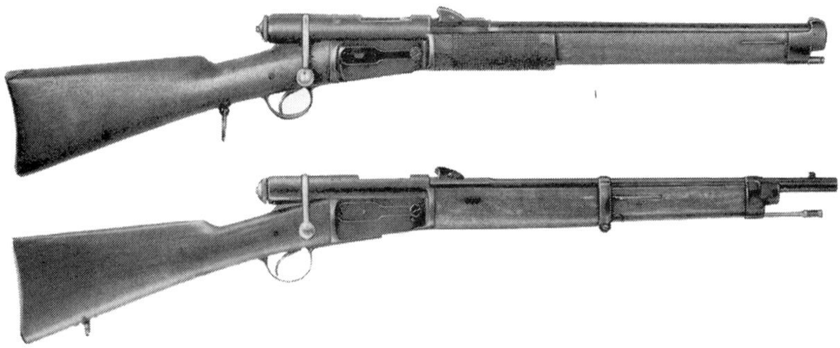

Fig. 1-46. Model 1878 Vetterli Carbines: top, Repeating Carbine for mounted troops; below, Repeating Carbine for Frontier Guards. James Wesson collection.

A total of 1,720 new carbines were manufactured at Eidgenössische Waffenfabrik, Bern from 1879 to 1885.

Model 1878 Vetterli Repeating Carbine - This model was very similar to the Model 1871 carbine for mounted troops but carried some of the improvements made to the Model 1878 series, refer to Figure 1-46. It was 36.6 inches long overall with an 18.5-inch-long barrel. The magazine held six cartridges. The conventional bolt handle with the ball tip was used and the cavalry sling ring attached to the bottom of the stock was retained. This carbine also retained the loading port cover. A new sight, designated the Model 1878 Rear Sight for Carbines, was introduced. It was a quadrant-style sight designed by Rudolf Schmidt and was graduated to 600 meters. Like the Model 1871 carbine for mounted troops, the forearm had a single step but only used the forward barrel band. The front sight was machined from the top of that barrel band. The shallow curved butt plate was retained. No bayonet was issued with this carbine.

Model 1878 Vetterli Carbine for Frontier Guards - Frontier guards patrolled the border on foot and horseback. As they were often in mountainous terrain, they required a short, handy weapon. The carbine was 37.2 inches long overall with a barrel 19.1 inches long. The forearm was not crosshatched and was held to the barrel and receiver with a key and two barrel bands. The rear barrel band also held the forward sling swivel. The forward barrel band was similar to that used on the rifle and had the "Mauser-style" bayonet lug on the right side. The loading gate cover was retained and the Schmidt quadrant rear sight graduated to 600 meters was used. The rear sling swivel was inletted into the bottom of the buttstock and the cavalry sling ring was not used. The conventional bolt handle with the ball end was also retained, refer to Figure 1-46.

Model 1881 Vetterli, Infantry Rifle

This was the last variation of the Vetterli rifle before the adoption of the Schmidt design at the end of the decade, see Figure 1-47. It was very similar to the Model 1878 with the exception of a redesigned trigger mechanism that provided a smoother action. The new Model

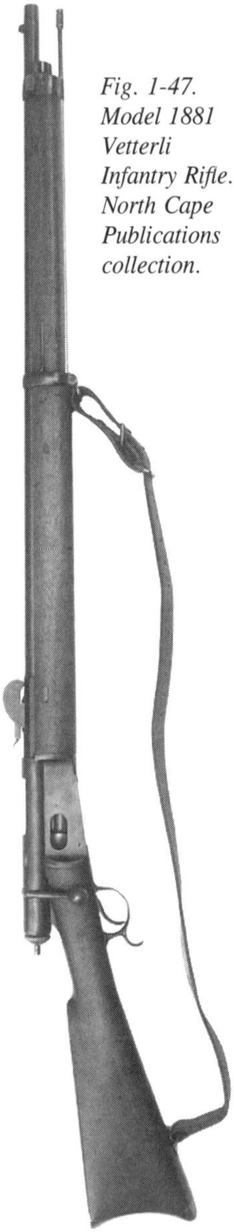

Fig. 1-47. Model 1881 Vetterli Infantry Rifle. North Cape Publications collection.

1881 quadrant rear sight was installed. It was calibrated for 1,600 meters. The rear sight base was stamped with range graduations of from 225 to 1,200 meters. The sight leaf was changed to include a leaf that telescoped in on itself. The extended slide was marked from 1,300 to 1,600 meters, refer to Figure 1-28.

The rifle was 51.9 inches long overall with a barrel 33.2 inches long. The magazine held twelve cartridges with an additional cartridge in the elevator. The rifle was equipped with a "Mauser-style" bayonet lug on the right side of the forward barrel band for the Model 1878 saw-toothed saber bayonet. The forend was smooth and had a single step for the rear barrel band. The sling swivels were mounted on bottom of the rear barrel band and the bottom of the buttstock. The deeply curved butt plate developed for the Model 1871 *Stutzer* rifle was used.

All 16,000 Model 1881 Vetterli Rifles were manufactured at Eidgenössische Waffenfabrik, Bern between 1881 and 1887.

Model 1881 Vetterli Short (*Stutzer*) Rifle

A so-called short-rifle version of the Model 1881 was also produced. It was 51.9 inches long overall with a 33.2-inch-long barrel, see Figure 1-48. The rifle did not use the loading gate cover. The reinforcing key was

Fig. 1-48. Model 1881 Vetterli Stutzer *Rifle. Photo courtesy of Swiss Ministry of Defense.*

used in the forend as well as a rear and forward barrel band. The forend was smooth, without cross-hatching. The rifle, like all *Stutzers,* was equipped with double-set triggers. The Model 1881 1,600-meter rear sight with the telescoping leaf was installed. The deeply curved butt plate developed for the Model 1871 *Stutzer* was also used. It was equipped with a "Mauser-style" bayonet stud on the right side of the forward barrel band for the Model 1878 saber bayonet with saw teeth. Like the Model 1878, the left side of the receiver was marked with the manufacturer's name and address, the serial number and the model year, refer to Figure 1-22. Magazine capacity was twelve.

NOTE: All Model 1878 and 1881 Vetterli rifles and carbines had a small hole drilled in the butt end of the stock. A spare rimfire firing pin wrapped in greased paper and tin foil was carried in the hole. The butt plate had to be removed to get at it.

Vetterli Rifle, Markings

Ordnance inspection, military and manufacturer's markings are described in detail in Chapter 5.

Chapter 2
The Schmidt-Rubin System

In 1886, the Swiss government ordered the military to establish a commission to investigate the development and adoption of a magazine repeating rifle. While the Vetterli had served well enough, its large 10.4 mm bullet and short effective range had made it obsolete as a frontline infantry rifle. The old bugbear of the tubular magazine, the changing point of balance as the magazine emptied, made it difficult to shoot accurately in rapid fire.

In 1881, Major, later Lieutenant Colonel, Eduard Rubin, director of the Thun munitions factory, had developed a new nine-millimeter lead bullet able to withstand higher pressures and velocities. The lead bullet was clad in a copper jacket, and the propelling charge was compressed black powder.

Over the next few years, the concept underwent further development. It was found that the bullet was most efficient in a rifle-length barrel when the diameter of the bullet was reduced to either 6.5 or 7.5 mm diameter. But pure copper jackets were difficult to apply at that time. Other jacket materials, including steel, were tested. An alloy of copper and zinc, called tombac, turned out to be the most cost effective. Experiments with bullet shape showed that a boat-tailed bullet provided the best accuracy and range, twice that of the old 10.4 mm black powder cartridge. The results of Lieutenant Colonel Rubin's work were impressive enough that the Commission directed their efforts toward finding a suitable magazine rifle for the new, small-caliber cartridge.

An interesting sidelight to the history of small-arms development both in Switzerland and the United States occurred in 1888. Lieutenant Colonel Rubin had also developed a magazine rifle that was an improvement over the Vetterli design. The Rubin Rifle was chambered for his new centerfire, reduced-caliber, high-velocity cartridge.

The Rubin rifle looked very much like the Vetterli rifle but had three locking lugs at the rear of the bolt and the firing pin was positioned for the centerfire cartridge. The barrel was 30.85 inches long with three grooves and lands, and the rifling twist was 1:10.625 inches.

The magazine was the fixed-box type holding seven cartridges. The magazine could be charged from a charger clip made of pressed paper and tin. The cartridge had a rimless brass, centerfire, tapered case. The 7.5 mm diameter bullet was copper-jacketed lead, round-nosed and weighed 217 grains. A charge of 69.48 grains of compressed black powder produced a velocity of 1,761 feet per second as measured fifty feet beyond the muzzle.

The Rubin Rifle was tested by the United States Army Ordnance Department at Springfield Armory between August and September 1888. The "Report of the Chief of Ordnance to the Secretary of War for Fiscal Year 1889" stated that the trial results of the Rubin Rifle and its ammunition were very satisfactory.

The Rubin Rifle was again entered in the U.S. 1892 Magazine Rifle Trials, as was the Model 1889 Schmidt Rifle as adopted by the Swiss Army. But both rifles lost to the Norwegian-designed Krag.

During this same period, Major Rudolf Schmidt, director of the Bern Federal Arsenal, had surveyed all the military rifles then in use. At the conclusion of his study, he had developed a design of his own that incorporated a variation of the Mannlicher straight-pull bolt and a steel box magazine based somewhat on the Lee pattern. The magazine could be fully charged with two tin and pasteboard chargers holding six cartridges each.

Major Rubin's cartridge design and charger clip was selected for the new rifle and designated the Model 1890. In its final form it was a rimless, centerfire, bottle-neck cartridge with a 211-grain, round-nose, boat-tail bullet. But compromises had to be made. Economics dictated that an iron-jacketed, lead, paper-patched bullet replace the very modern boat-tail, tombac-clad bullet that Major Rubin preferred. The charge was probably *rauchschwache,* which provided a muzzle velocity of between 1,903 and 1,968 feet per second.

NOTE: The exact composition of *rauchschwache* is not known to the author. Eric Timrud, a knowledgeable collector of Vetterli and Schmidt rifles and ammunition, disassembled a Model 1890 cartridge. When the powder charge was ignited, there was no sulphur dioxide odor characteristic of a black powder or "semi-smokeless" based charge.

The resulting rifle was long, heavy and somewhat awkward to carry. The straight-bolt system gave up the mechanical advantage of positive extraction for rapidity of manipulation. The receiver was very long, much longer than needed as successive modifications would show. The bolt's locking lugs were mounted on the rear of the bolt sleeve to avoid fouling with powder residue, and they rotated into matching locking recesses in the back of the receiver. The result was a bolt assembly capable of handling the 37,500 psi pressures generated by the Model 1890 cartridge, but not much more. The Model 1889 Schmidt rifle was accurate for its time, and sturdy, which is what armies demand first and foremost. Table 2-1 summarizes all Schmidt rifle and carbine models adopted by the Swiss government between 1889 and 1956. Tables 2-2 and 2-3 summarize the issue and certain other types of ammunition of interest for the Schmidt rifles during this time period.

Table 2-1
Repeating Rifle System, Schmidt Model

Model	Type	Chief Identifying Characteristics
1889	Infantry Rifle	51.3 inches overall, 30.7-inch barrel, 9.3-inch long receiver, dual rear-mounted locking lugs mounted at rear of the bolt locking sleeve, 12-round magazine, bayonet mount
1889/96	Infantry Rifle	51.2 inches overall, 30.7-inch barrel, 8.46-inch long receiver, dual rear-mounted locking lugs moved forward 0.79 inch at the rear of bolt locking sleeve, 12-round magazine, bayonet mount

Table 2-1, cont. Repeating Rifle System, Schmidt Model		
Model	**Type**	**Chief Identifying Characteristics**
1889/00	Short Rifle	43.5 inches overall, 23.3-inch barrel, large dual rear-mounted locking lugs as in the M1889/96, six round magazine, bayonet mount
1905	Cavalry Carbine	42.1 inches overall, 21.6-inch barrel, six-round magazine, no bayonet mount
1911	Infantry Rifle	51.6 inches overall, 30.7-inch barrel, dual front-locking lugs on bolt sleeve, six-round magazine
Karabiner 1911	Carbine	43.4 inches overall, 23.3-inch barrel, dual front-locking lugs on bolt sleeve, six-round magazine
1896/11	Infantry Rifle	51.3 inches overall, 30.7-inch barrel, 8.46-inch-long receiver in its original form. Almost all Model 1889/96s were converted to the Model 1896/11 configuration
Karabiner 31	Rifle	43.6 inches overall, 25.7-inch barrel, dual front-locking lugs on bolt sleeve, six-round magazine, integral with trigger guard

Table 2-2 The Swiss Schmidt Ammunition (Source: Handfeuerwaffen Gradzug-Systeme)				
Designa-tion	**Dimen-sions (mm)**	**Bullet Type/Weight**	**Propellant**	**Velocity (fps)**
GP 1890 (1)	7.5 x 53.5	Copper-plated iron cap, round nose, paper patched, lead .321-inch diam-eter (2)	29 grains, unknown propellant	Rifle: 1,903-1,968; Carbine: 1,837

Table 2-2, cont. The Swiss Schmidt Ammunition (Source: Handfeuerwaffen Gradzug-Systeme)				
Designa-tion	**Dimen-sions (mm)**	**Bullet Type/Weight**	**Propellant**	**Velocity (fps)**
GP 1890/03	7.5 x 53.5	Copper-plated iron cap, round nose, paper patched, lead .321-inch diameter (2)	29 grains, unknown propellant	Rifle: 1,903-1,968; Carbine: 1,837
GP 1890/23	7.5 x 54.5	Cupronickel (3) jacketed, round nose, .308 inch at case mouth flaring to .310 inch at base (4)	smokeless	1,953
GP 1911	7.5 x 55	Cupronickel (3) plating on steel jacket over lead core, boat tail, spire point, .308-inch diameter	smokeless	M1911 Rifle: 2,500 K31 Carbine: 2,559

1. GP, abbreviation for "Gewehr Patrone" (rifle cartridge)
2. Diameter includes paper patch
3. 20% nickel, 80% copper in jacket material
4. Measured by Eric A. Timrud

Table 2-3 Some Other Swiss Schmidt Ammunition		
Designation	**Dimensions (mm)**	**Identification**
Kadett	7.5 x 53.6	Cadet, blank
GP 1890	7.5 x 53.6	Blank, red wooden bullet

Table 2-3, cont. Some Other Swiss Schmidt Ammunition		
Designation	**Dimensions**	**Identification**
GP 1890	7.5 x 53.6	Dummy cartridge
GP 1911	7.5 x 55.6	Blank, Green wooden bullet
GP 1923	7.5 x 54.6	Cadet, live cartridge
GP 1924	7.5 x 55	Tracer, red base, slightly flattened bullet tip
GP 1928	7.5 x 55	Armor-piercing, violet base, cupronickel or tombac jacket
GP 1944	7.5 x 55	Grenade-launching blank, zinc-washed, elongated case with crimped mouth, large knurled ring, 3.035 inches long
GP 1944/48	7.5 x 55	Grenade-launching blank, elongated cartridge, wooden ball crimped in mouth, 2.661 inches long
GP 1955	7.5 x 55	Blank, round nose plastic bullet (green but also red and white), diameter .297 inch, case length 2.175 inches
GP 1958	7.5 x 55	Blank, round nose green plastic bullet, diameter 0.298 inch, case length 2.177 inches

Repeating Rifle, Model 1889

First introduced to the Swiss Army in 1890, this long rifle and its twelve-round box magazine was eagerly accepted by Swiss troops, see Figure 2-1. The new, "small"-caliber Model 1889 rifle firing the Model 1890 7.5 x 53.5 mm rimless cartridge was far more accurate than the old black powder Vetterli and had more than twice the range.

The rifle's length, its long receiver and its straight-pull bolt with the large ring for manually cocking the bolt (see Figure 2-2) make it instantly recognizable. Its stacking swivel, a narrow-diameter rod with

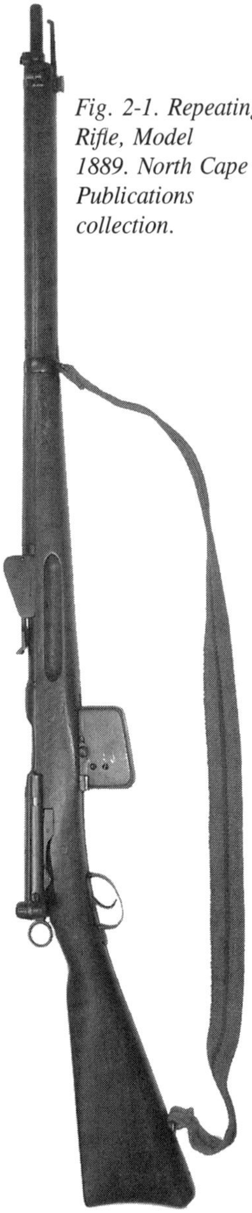

Fig. 2-1. Repeating Rifle, Model 1889. North Cape Publications collection.

a curled tip mounted on the front barrel band, was the first of its type to be used. The curl served two purposes; it provided a more secure hold when three rifles were "stacked" into a tripod by interweaving the stacking rods, and it reduced the chances of brush and twigs catching between the rod and barrel, see Figure 2-3.

The other distinctive features are the massive straight-pull bolt with the bolt handle projecting at 90 degrees to the bolt body, and its wooden-knobbed grasping handle. The twelve-round box magazine projected two inches below the bottom of the stock. A magazine cutoff feature was incorporated. By pushing down on a switch on the side, the magazine was lowered enough that the bolt face did not make contact with the top cartridge on the magazine follower. A single cartridge could be inserted into the chamber and the bolt closed.

While the Model 1889 today seems to be an extremely long and heavy rifle for the late 19th century, it was only one-half inch longer than the Model 1871/86 Mauser and three-quarter inch shorter than the Portuguese Kropatschek.

The Model 1889 was chambered for the Gewehr Patrone (GP) 1890 7.5 x 53.5 mm rimless cartridge with a paper-patched, round nose, copper-plated, iron-jacketed lead, 213-grain bullet with a diameter of 0.321 inch, including the patch. The greased paper jacket lubricated the bullet

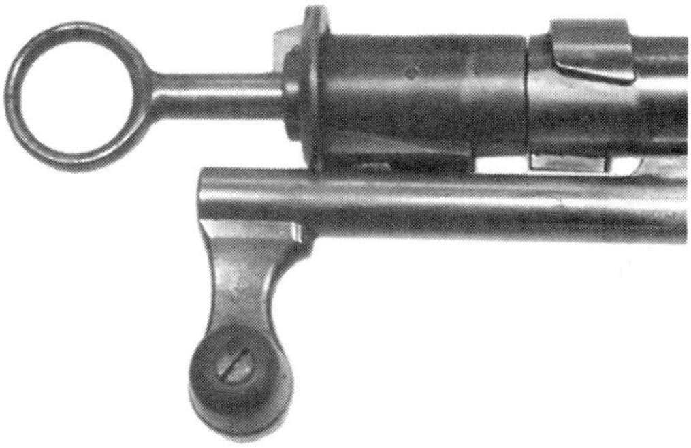

Fig. 2-2. *Manual cocking ring on the Model 1889 Rifle. This was a characteristic feature of all so-called Schmidt-Rubin rifles.*

on its passage down the bore and helped reduce fouling from the powder charge. The priming compound was mercury fulminate-based and so did not produce corrosion in the barrel.

Because the three lands and grooves are of equal width, they each occupy one sixth of the barrel's diameter. The rifling had a right-hand twist of 1:10.63 inches. The diameter of the bore measured along the tops of each groove is 0.296 inch with grooves 0.007 inch deep. The diameter of the bore measured from the top of one land to the bottom of the opposite groove is 0.303 inch. The diameter of the bore from the bottom of each groove is 0.310 inch, Figure 2-4.

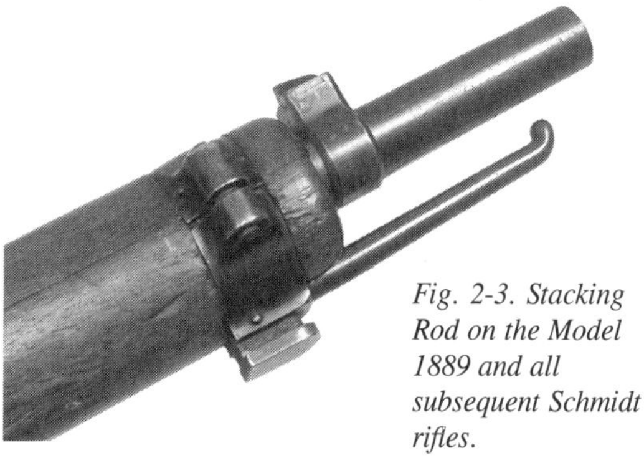

Fig. 2-3. *Stacking Rod on the Model 1889 and all subsequent Schmidt rifles.*

The GP 1890 cartridge contained a nominal 29-grain charge of semi-smokeless powder of a composition unknown to the author, and developed a muzzle velocity of between 1,903 and 1,968 fps in the rifle-length barrel. The GP 1890 cartridge powder loading varied between 27 and 31 grains depending on the quality of the powder.

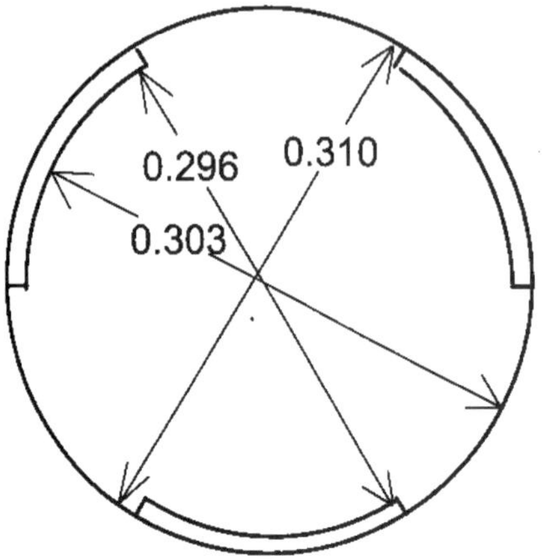

Fig. 2-4. Model 1889 rifle bore dimensions, in decimal inches.

In 1903, a new, less erosive primer was introduced and the cartridge was designated the GP 1890/03. The bullet's velocity remained the same. The "iron jacket" of the GP 1890 and 1890/03 cartridge is often erroneously referred to as "steel" when in fact it was made of low-carbon iron, and then copper-plated. The GP90/03 remained 53.5 mm long. When fired in the Model 1889 or Model 1889/96/00 rifles, both the GP 1890 and GP 1890/03 developed a breech pressure of 37,000 psi.

In 1923, the old GP 1890/03 cartridge was redesigned slightly. The round nose bullet shape was retained but its weight was lowered to 190 grains. The bullet was given a new jacket containing 80% copper and 20% nickel. A smokeless charge was substituted for the semi-smokeless powder charge used in the previous two cartridges. The new cartridge was 54.5 mm long, 1 mm longer than the GP90 or GP90/03, but the chamber of the Model 1889 was long enough to handle it safely. The GP90/23, as it was designated, produced the same chamber pressure as the GP90 or GP90/03 and provided the *Landsturm* who were still equipped with unaltered Model 1889 rifles with a full metal jacketed cartridge. The *Landsturm* were the inactive reserve of the Swiss Army.

Some 212,000 Model 1889 rifles were manufactured by Eidgenössische Waffenfabrik, Bern beginning in 1891 and ending in 1897. See Table 2-4.

NOTE: While the GP11 cartridge was authorized by the Swiss Army to be used in the pre-Model 1911 rifles and carbines in "wartime" only, the modern shooter must not do so as the weaker locking lugs on these rifles may fail with the more powerful cartridge, which is described below.

Table 2-4 Repeating Rifle, Model 1889	
Overall length	51.3 inches
Barrel	30.7 inches long, 3 grooves, right-hand twist, 1:10.63 (inches) rifling
Weight, unloaded, no bayonet	10.69 lbs
Action Type	Schmidt straight pull, rotating bolt collar with two opposed locking lugs at rear
Caliber	GP90, GP90/03 (7.5 x 53.5), GP90/23 (7.5 x 54.5)
Chamber Pressure (maximum)	37,000 psi
Magazine Capacity	12
Rear Sight Type	Quadrant, adjustable for elevation only, 2,000 meters, in 100-meter increments
Total Military Production	212,000

Repeating Rifle, Model 1889/1896

The Model 1889/1896 was identical in most respects to the Model 1889 with the exception of its receiver length, see Figure 2-5. The bolt locking lugs were made slightly larger and moved 0.79 inch forward for increased strength. This allowed the receiver to be shortened by 0.79 inch. The magazine cutoff feature was retained.

Some 137,050 Model 1889/1896 rifles were manufactured at Eidgenössische Waffenfabrik, Bern from 1895–1912. See Table 2-5.

Fig. 2-5. Repeating Rifle, Model 1889/1896.

Table 2-5 Repeating Rifle, Model 1889/1896	
Overall Length	51.2 inches
Barrel	30.7 inches long, 3 grooves, right-hand twist, 1:10.63 (inches) rifling
Weight, unloaded, no bayonet	9.92 lbs
Action Type	Schmidt straight pull, rotating bolt collar with two opposed locking lugs at the mid-position
Caliber	GP90, GP90/03 (7.5 x 53.5), GP90/23 (7.5 x 54.5)
Chamber Pressure (maximum)	37,000 psi
Magazine Capacity	12
Rear Sight Type	Quadrant, adjustable for elevation only, 2,000 meters in 100-meter increments
Total Military Production	137,050

Repeating Rifle, Short, Model 1889/1900

After ten years of service, it had become clear that the twelve-round magazine was unnecessary and the 51.2-inch Model 1889/96 rifle was far too long for efficient service with mobile, support and rear-echelon troops, refer to Figure 2-6. The magazine protruded too far below the bottom of the stock and hampered the soldier when firing in the

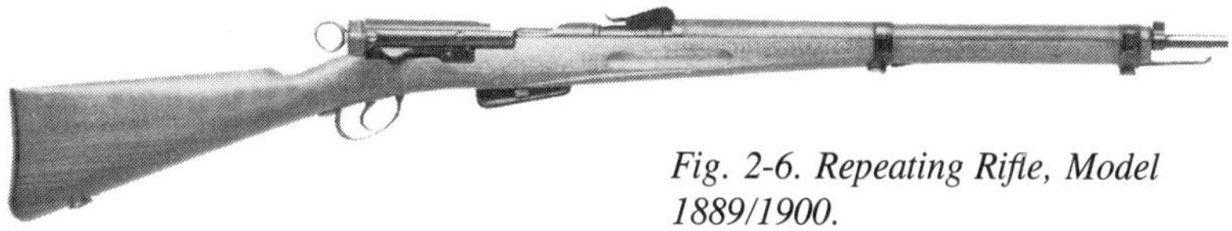

Fig. 2-6. Repeating Rifle, Model 1889/1900.

prone position; the rifle's length made movement in heavy brush or forested areas difficult. And it was difficult to manage well on horseback or bicycle. A series of tests of the clip-loading feature showed that it was just as fast to load six cartridges twice using one clip each time as twelve cartridges once using two clips. The magazine capacity was therefore reduced to six rounds and the magazine redesigned as described below in the section on magazines. The new six-round magazine enabled the soldier to fire from the prone position and eased the problem of misfeeds that occasionally troubled the long magazine. The barrel was also shortened by 7.4 inches, which reduced the overall length by 7.7 inches to 43.5 inches. The action remained unchanged from the Model 1889/1896 and the new rifle (see Figure 2-6) weighed 7.9 lbs, a welcome 2.02 pounds less than its predecessor.

Mounted troops as well as other branches of the service such as the artillery, signal and bicycle units, which had complained of the Model 1889's length, were issued the new Model 1889/1900 short rifle. Records indicate that 18,750 were manufactured from 1901 to 1911 by Eidgenössische Waffenfabrik at Bern, Switzerland. See Table 2-6.

Table 2-6 **Repeating Rifle, Model 1889/1900**	
Overall length	43.5 inches
Barrel	23.3 inches long, 3 grooves, right-hand twist, 1:10.63 (inches) rifling
Weight, unloaded, no bayonet	7.9 lbs

Table 2-6, cont. Repeating Rifle, Model 1889/1900	
Action Type	Schmidt straight pull, rotating collar with two opposed locking lugs at the rear
Caliber	GP90, GP90/03 (7.5 x 53.5), GP90/23 (7.5 x 54.5 when modified)
Chamber Pressure (maximum	37,000 psi
Magazine Capacity	6
Rear Sight Type	Quadrant, adjustable for elevation only, 1,200 meters in 100-meter increments
Total Military Production	18,750

Repeating Carbine, Model 1905

The short rifle, Model 1889/1900, was so successful that Ordnance followed through with the Model 1905 carbine, see Figure 2-7. This was the first Schmidt design to be made specifically for pioneer (engineer) and mounted troops. The Model 1905 carbine replaced the unpopular Model 1893 carbine, which had used the Mannlicher straight-pull bolt system and clip-fed magazine and was manufactured at Steyr, Austria.

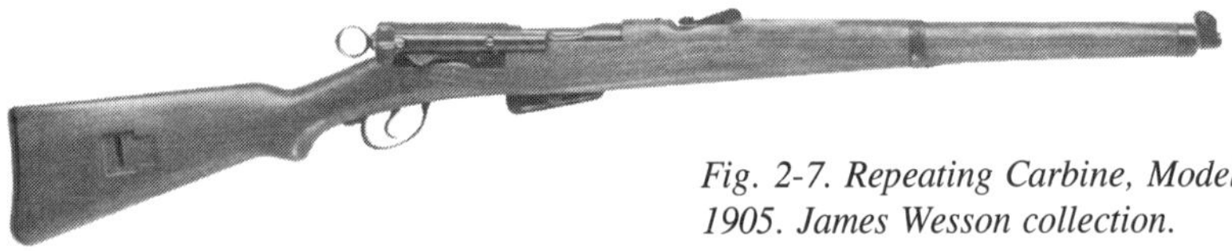

Fig. 2-7. Repeating Carbine, Model 1905. James Wesson collection.

The Model 1905 carbine was basically the Model 1889/1900 rifle shortened to 42.1 inches overall. The barrel was 21.6 inches long. The rear sling swivel was inletted into the left side of the stock and consisted of a steel bar bolted vertically top and bottom above the "V"-shaped cutout. The forward sling swivel was mounted on the left side of the rear barrel band. The carbine was stocked to the muzzle and had no provision for a bayonet. A total of 7,900 were manufactured between 1906 and 1911 at Eidgenössische Waffenfabrik, Bern, see Table 2-7.

Table 2-7 **Repeating Carbine, Model 1905**	
Overall Length	42.1 inches
Barrel	21.6 inches long, 3 grooves, right-hand twist, 1:10.63 (inches) rifling
Weight, unloaded, no bayonet	7.94 lbs
Action Type	Schmidt straight pull, rotating bolt collar with two opposed locking lugs at front of collar
Caliber	GP90, GP90/03 (7.5 x 53.5), GP90/23 (7.5 x 54.5 when modified)
Chamber Pressure	37,000 psi
Magazine Capacity	6
Rear Sight Type	Quadrant, adjustable for elevation only, 1,500 meters in 100-meter increments, 200-meter fixed sight
Total Military Production	7,900

Repeating Rifle, Model 1911

This redesign of the Schmidt rifle was driven by the desire to adopt a new cartridge, the Model 1911 for infantry use. The cartridge was 7.5 x 55 mm long overall with a 174-grain spire-point, boat-tail bullet

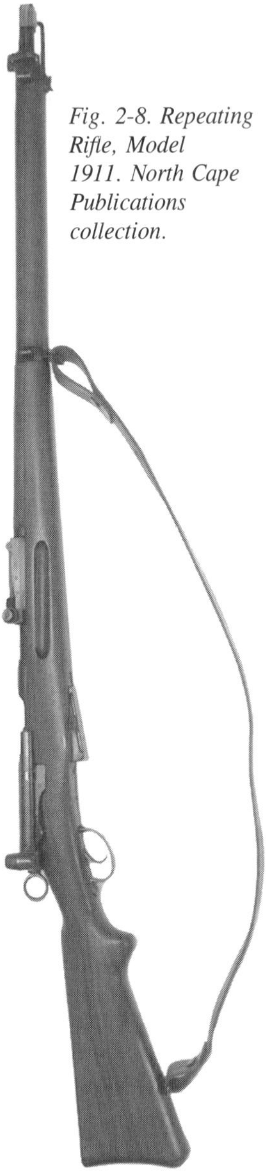

Fig. 2-8. Repeating Rifle, Model 1911. North Cape Publications collection.

0.308 inch in diameter. The bullet had a lead core with a cupronickel jacket (80% copper and 20% nickel). True smokeless powder was used for the first time in a Swiss general-issue military cartridge. The chamber pressure was increased to 45,000 psi and the bullet's muzzle velocity to 2,500 fps, see Figure 2-8.

NOTE: The GP 1911 cartridge exceeds the rated chamber pressure of pre-1911 rifle and carbine models by 8,000 psi. Even though the Swiss Army allowed its use in pre-1911 rifles as a war emergency measure, it should not be loaded and fired in the Models of 1889, 1889/96, 1889/00 and 1905.

The outward changes to the Model 1911 included a new stock with a pronounced pistol grip in somewhat the same shape as found on the Lee-Enfield SMLE, see Figure 2-9. A six-round magazine without a magazine cutoff, proven in the Models of 1889/00 and 1905, replaced the twelve-round magazine used in the Model 1889 and Model 1889/96. The six-round magazine protruded only 0.6 inch below the trigger guard plate and did not interfere with firing from the prone position. A new rear sight and a slightly improved front sight replaced the old Model 1889 sights. The new rear sight was of the tangent variety and patterned after those used on the Danish (Model 1889) and American (Model 1898) Krag rifles and carbines. The base had a convex curve, and the elevation slide moved along this curve to raise or lower the leaf to change the elevation, see Figure 2-10.

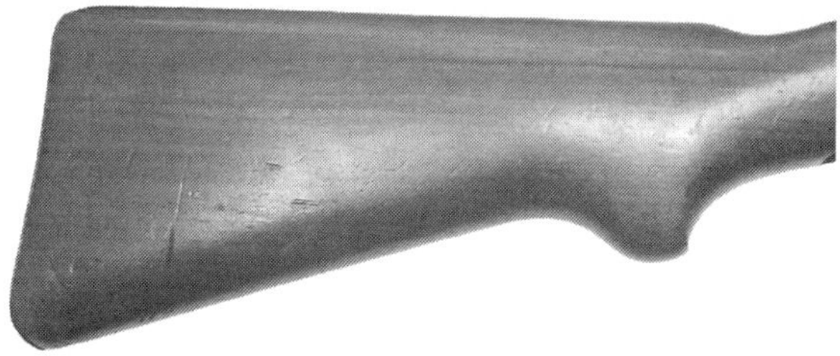

Fig. 2-9. Model 1911 buttstock with pistol grip.

The receiver was changed very little in spite of the use of the new higher-pressure GP11 cartridge. If you examine the bottom of the Model 1889/96/00/05 receivers and compare them to the Model 1911, you will see that only the cuts at either end of the magazine well, to accommodate the reinforcing band of the earlier magazines, were eliminated. Otherwise, except for markings, the receivers are basically the same.

Fig. 2-10. Model 1911 tangent rear sight.

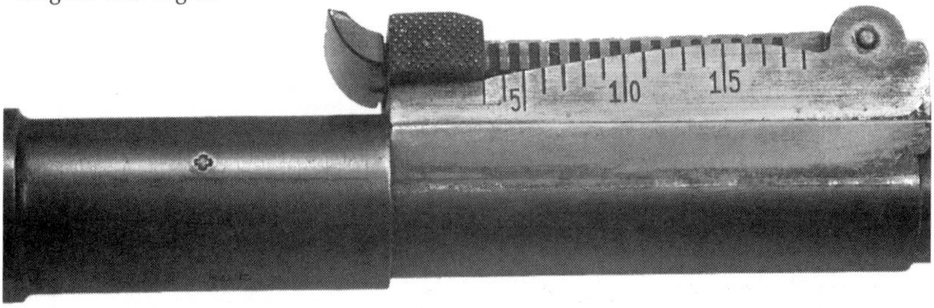

Production of the Model 1911 rifle began in 1912 and ended in 1919 after 126,700 had been manufactured. See Table 2-8.

Table 2-8 Repeating Rifle, Model 1911	
Overall Length	51.6 inches
Barrel	30.7 inches long, 4 grooves, right-hand twist, 1:10.63 (inches) rifling
Weight, unloaded, no bayonet	10.3 lbs
Action Type	Schmidt straight pull, rotating bolt collar with dual front-locking lugs on bolt sleeve
Caliber	GP11 (7.5 x 55 mm)
Chamber Pressure (maximum)	45,000 psi
Magazine Capacity	6
Rear Sight Type	Tangent sight, leaf graduated to 2,000 meters in 100-meter increments
Total Military Production	126,700

Repeating Carbine, Model 1911

The new Model 1911 carbine replaced both the Model 1905 carbine and the Model 1889/1900 short rifle, see Figure 2-11. It was only one tenth of an inch shorter than the Model 1889/1900 short rifle but was capable of firing the newer, more powerful GP 1911 cartridge. The new tangent rear sight was also used on the carbine but graduated for the shorter barrel. The front sight was a new design protected by ears and much easier to "zero" at the factory. It was also equipped to accept the long Model 1911 bayonet with a saw back, which was issued to pioneer troops for cutting through the wooden stakes holding up barbed wire entanglements, clearing brush, etc.

NOTE: The Model 1889/99/18 bayonets can be mounted on the Model 1911 carbine. But the bores of these bayonets are too large to fit the barrel securely. Only the Model 1911 bayonet with its smaller bore was authorized for use on the Model 1911 carbine.

The Model 1911 carbine was manufactured from 1914 to 1933. A total of 212,100 are believed to have been made, according to serial number records. See Table 2-9.

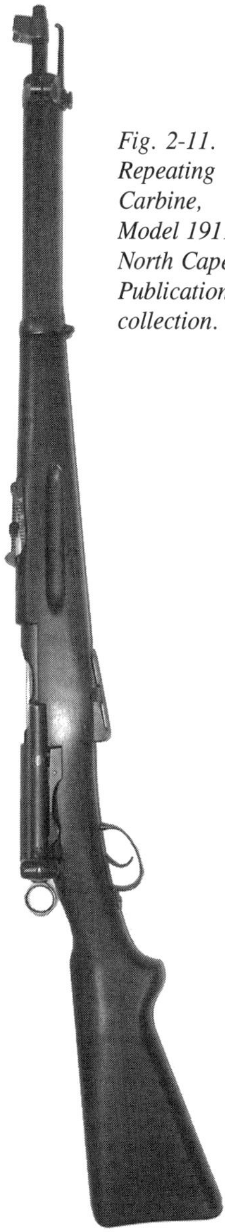

Fig. 2-11. Repeating Carbine, Model 1911. North Cape Publications collection.

Repeating Rifle, Model 1896/1911

It was a hallmark of 19th century military organizations everywhere that nothing was ever discarded until every possible bit of use was wrung from it, particularly small arms. With the advent of the Model 1911 rifle and its new, spire-point, .308-caliber bullet and a modified shoulder, the Swiss were anxious to rearm completely. But "new rifles" were very expensive to produce and the capacity of the weapons factories and their subcontractors was limited. So they turned to the stockpile of 137,050 obsolete Model 1889/1896 rifles in armories and in the hands of rear-echelon and support troops. They were modified to fire the new GP 1911 cartridge.

The rifles were easily rebuilt. A pistol grip was cut into the old stock or, if the original stock was not reusable, a new one replaced it, Figure 2-12. The square contour

Table 2-9 Repeating Carbine, Model 1911 (K11)	
Overall Length	43.4 inches
Barrel	23.3 inches long, 4 grooves, right-hand twist, 1:10.63 (inches) rifling
Weight, unloaded, no bayonet	8.85 lbs
Action Type	Schmidt straight pull, rotating bolt collar with dual center-mounted locking lugs
Caliber	GP11 (7.5 x 55 mm)
Chamber Pressure (maximum)	45,000 psi
Magazine Capacity	6
Rear Sight Type	Tangent sight, leaf graduated to 1,500 meters in 100-meter increments
Total Military Production	212,100

to the top of the buttstock remained as did the original butt plate. A new trigger guard plate was manufactured with a swell to the right side near the front to fill the hole in the stock left by the removal of the magazine catch. The original M1889/96 bolt, barrel, sights and magazines

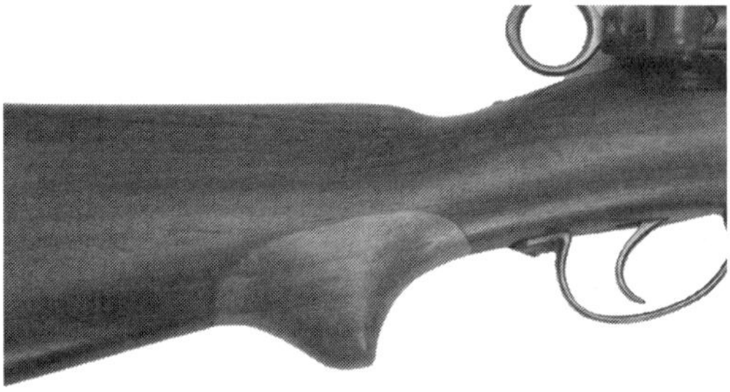

Fig. 2-12. Those Model 1889/96 rifles modified for the GP11 cartridge usually have a pistol grip inletted into the original stock.

were retained. The Model 1896/1911 is functionally the same as the newly manufactured Model 1911 rifles.

It is thought that the total of rifles so modified nearly approached the original production number of 137,050. See Table 2-10.

Table 2-10 Repeating Rifle, Model 1896/1911	
Overall Length	51.3 inches
Barrel	30.7 inches long, 4 grooves, right-hand twist, 1:10.63 (inches) rifling
Weight, unloaded, no bayonet	9.94 lbs
Action Type	Schmidt straight pull, rotating bolt collar with two opposed locking lugs at front of collar
Caliber	GP11 (7.5 x 55 mm)
Chamber Pressure (maximum)	45,000 psi
Magazine Capacity	6
Rear Sight Type	Tangent sight, leaf graduated to 2,000 meters in 100 meter increments
Total Military Production	137,050 (±)

Repeating Rifle (Karabiner), K31

In the Model 1911 the locking lugs on the bolt appear to have been moved to the front of the bolt body. In fact, the locking lugs remained in the same position on the bolt collar, but the bolt body was shortened considerably to fit within the bolt collar. This allowed the new receiver

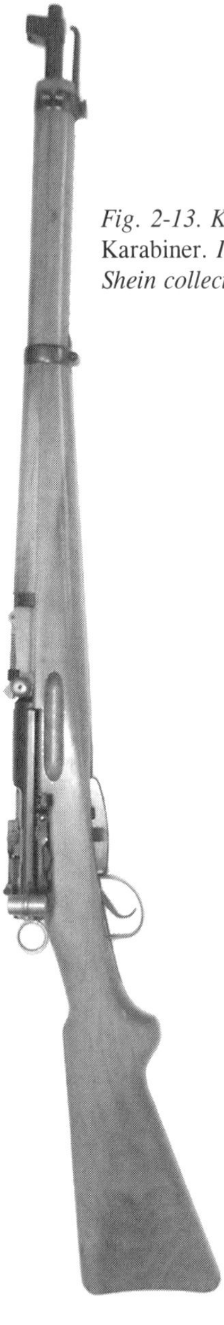

Fig. 2-13. K31 Karabiner. *Ian Shein collection.*

to be shortened by 1.9 inches over the Model 1911 rifle receiver.

NOTE: The K31 should not be considered a true Schmidt design. A Major Furrer is acknowledged to have redesigned the Model 1911. But as in the original Schmidt design, the bolt does not rotate; only the bolt collar.

The new shorter bolt and receiver also made it possible to move the magazine 2.5 inches to the rear which improved the balance of the weapon and simplified manufacture. The barrel length was increased by 2.4 inches over that of the Model 1911 carbine. But even with the longer barrel, the new rifle was only 0.2 inch longer than the Model 1911 carbine, see Figure 2-13 (43.6 vs 43.4 inches).

The new short rifle was termed a carbine (*Karabiner* in German, *le mousqueton* in French) and it was 1.45 lbs lighter than the Model 1911 rifle. Weight is the infantryman's second-worst enemy. Under the strain of combat, excess weight reduces the infantryman's ability to move quickly and quietly and exhausts him faster. The new, lighter rifle therefore proved very popular immediately.

The K31 can be readily identified by its pistol-grip stock, a magazine adjacent to the front of the trigger guard, a shorter receiver, a hand guard that ends at the rear sight and aluminum knobs on the bolt handle instead of bakelite.

The K31 rifle gradually replaced the

Model 1911 long rifle and carbine after its introduction and served with Swiss forces until it was replaced in turn with the *Sturmgewehr* 57 select-fire rifle beginning in 1958.

All 528,230 K31s were manufactured at Eidgenössische Waffenfabrik in Bern, Switzerland, between 1933 and 1963. An additional 600 K31s were manufactured between 1968 and 1972.

One hundred rifles between serial numbers 249,047 and 249,146 were reported to have been manufactured in 1955 and sent to the Papal Swiss Guards that provide the Papal security detail for Vatican City. See Table 2-11.

Table 2-11 **Repeating Rifle, Karabiner 1931**	
Overall Length	43.6 inches
Barrel	25.7 inches long, 4 grooves, right-hand twist, 1:10.63 (inches) rifling
Weight, unloaded, no bayonet	8.85 lbs
Action Type	Straight pull, non-rotating bolt with rotating bolt collar with two opposed locking lugs
Caliber	GP11 (7.5 x 55 mm)
Chamber Pressure (maximum)	45,000 psi
Magazine Capacity	6
Rear Sight Type	Tangent sight, leaf graduated to 1,500 meters in 100-meter increments
Total Military Production	528,230 + 600 + 100

Kadett Rifle, Model 1897

A shorter, lighter version of the Schmidt rifle was developed and provided for the use of cadets in military schools, see Figure 2-14.

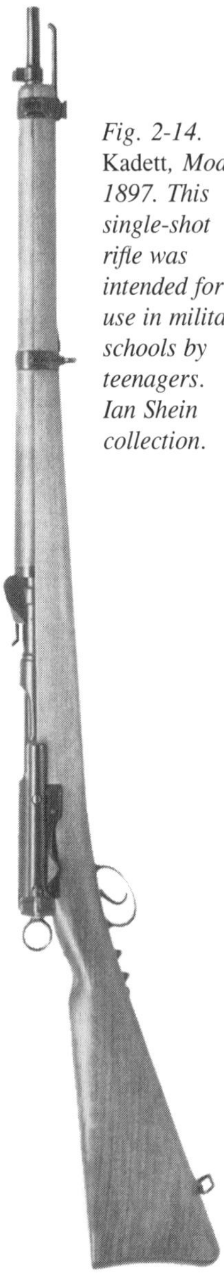

Fig. 2-14. Kadett, *Model 1897. This single-shot rifle was intended for use in military schools by teenagers. Ian Shein collection.*

The Model 1897 replaced the old Model 1870 Vetterli cadet rifle.

The new cadet rifle was 43.5 inches long overall, had a barrel 23.3 inches long and weighed 7.7 pounds unloaded and without a bayonet or sling. The rifle was designed as a single shot and so lacked the magazine system of the infantry rifle. But the bolt mechanism was very similar to that of the Model 1889/96 infantry rifle with the same, large manual cocking ring. The trigger guard had two finger rests. Two barrel bands were used. The forward barrel band was similar to that found on the Model 1889 and 1889/96 rifles, with stacking swivel and bayonet mount. The rear barrel band was equipped with a sling swivel, and the rear sling swivel was mounted on the bottom of the buttstock, ahead of the butt plate. The butt plate was similar to that used on the Model 1889/96 rather than the metal cap used on the Model 1889. The *Kadett* rifle used the Model 1889 knife bayonet.

Repeating Carbine, Model 1893 (Mannlicher)

The Model 1889 repeating rifle based on the Schmidt system had only been in service for two years when the Ordnance Department decided to adopt a carbine for mounted troops based on the straight-pull, revolving bolt system devised by Ferdinand Mannlicher, see Figure 2-15. The system was adopted from the Austrian Model 1888–90 carbine. It was chambered for the Swiss GP 1890 7.5 x 53.5

mm cartridge and had a six-round magazine based on Schmidt's version of the Lee box magazine but without the magazine cutoff feature of the Model 1889 rifle.

The carbine was 40.1 inches long overall with a 21.7-inch barrel and weighed 6.83 pounds unloaded. The carbine was stocked to the muzzle with a full-length hand guard that stretched from the front of the receiver ring to the front sight. Two barrel bands were used. The carbine was not equipped for a bayonet.

The bolt was very similar to the bolt used on the Austrian Model 1888–90 carbine. In place of the Schmidt manual cocking ring, a turned-up thumbpiece was used. The front sight was protected by ears on either side. The bolt handle projected at a 90-degree angle.

Fig. 2-15. Repeating Carbine, Model 1893. This carbine was actually a Mannlicher straight-pull turn bolt rather than a Schmidt straight-pull design. Photo courtesy of Swiss Ministry of Defense.

Chapter 3
The Schmidt Rifle and Carbine, Part-by-Part

The Schmidt rifle system was developed and used by the Swiss Armed Forces for sixty-eight years, and it remained in service as a second-line weapon for several years afterward. While it may not have been an optimum design in battle rifles, it suited the Swiss citizen soldier very well. The rifle was rugged and accurate and well able to stand the abuse of Swiss summers and winters.

In its sixty-eight year service history, only three major models were evolved, although two variations of the basic Model 1889 rifle pointed the way to the second, the Model 1911. The changes were made with an economy of design and production change that could only be achieved by the industrious Swiss. Many parts of the rifle survived intact without a single change, or else only a change in a type of screw or pin.

The following sections provide a part-by-part examination of the three major Schmidt rifle and carbine models, those of 1889, 1911 and the K31 Karabiner. Note that *Karabiner* (Carbine) means a short rifle.

NOTE: As parts of the Model 1889, Model 1889/1896, and Model 1889/1900 and, to a lesser extent, the Model 1911 were manufactured by a number of different subcontractors, minor variations in non-critical parts will be observed from company to company. As an example, barrel bands, stacking swivel rods, and the thickness of butt plates, among others, will vary slightly from those dimensions given below.

Stock Assembly, Model 1889 Rifle

Original Model 1889 stocks were made of European walnut, see Figure 3-1. The stock was made in one piece, was 49.13 inches long and

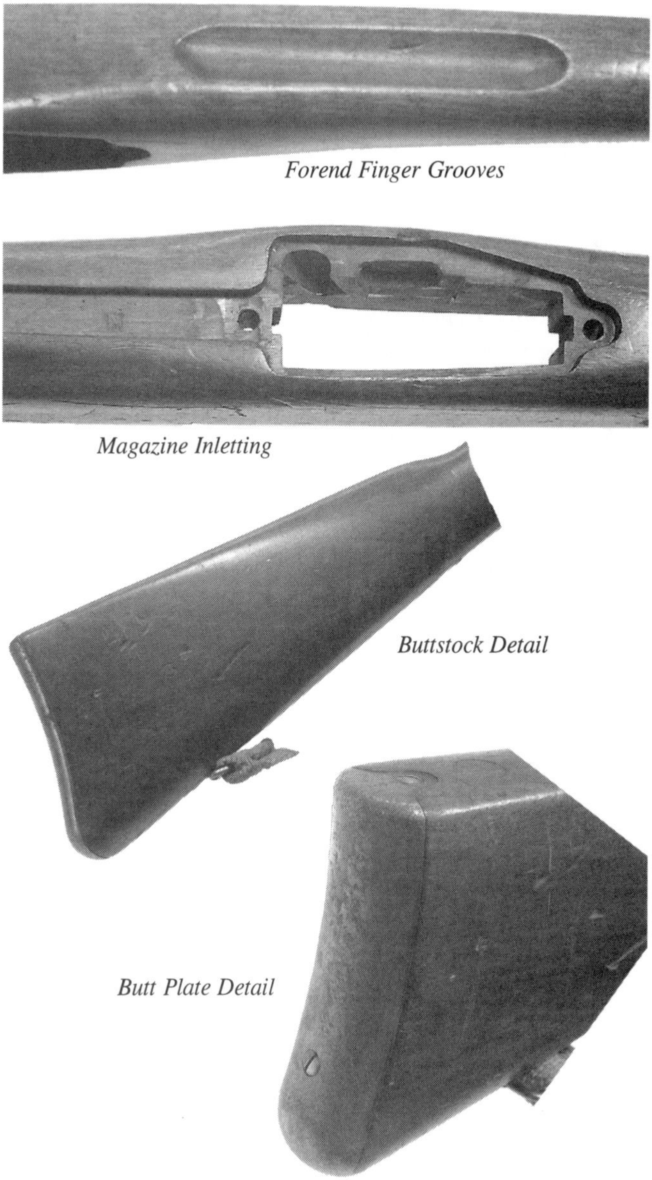

Fig. 3-1. Model 1889 Schmidt rifle stock details.

extended to within 2.75 inches of the muzzle. The forend had a 5.25-inch-long finger groove on either side below the rear sight. The comb was rudimentary and the butt end of the stock had a shallow curve. The butt plate was a heavy steel cap with a short upper tang, fastened to the butt with two wood screws. The Swiss Ordnance Department did not include either lightening channels in the buttstock or a butt trap in the butt plate.

The barrel channel was inletted for the shape of the barrel and the metal barrel collar that served to free-float the barrel and which was partly responsible for the rifle's accuracy, see Figure 3-2.

Two holes were drilled in the buttstock for the butt plate screws.

NOTE: All stock measurements are made from the top of the forend or forearm to the toe of the stock.

Stock Assembly, Model 1911 Rifle

The Model 1911 rifle stock assembly differed in only two particulars from the Model 1889 rifle stock. 1) It was 49 inches long, and 2) it lacked the inletting around the trigger guard plate seat for the magazine cutoff. Instead, that same area was bulged outward to allow room for the magazine release switch. The inletting for the magazine release was closed at the top, see Figure 3-3, arrow.

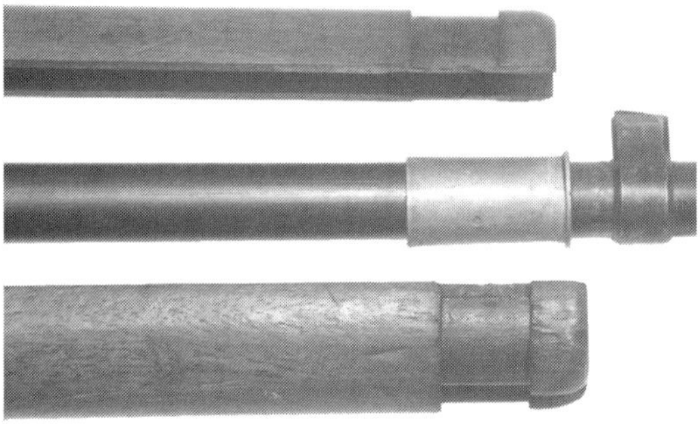

Fig. 3-2. Model 1889 barrel, barrel collar and hand guard. The barrel collar free-floated the barrel and contributed to the rifle's accuracy.

The butt cap used on the Model 1889 was replaced with a steel butt plate, which reduced the amount of hand work necessary to fit the butt cap. The finger grooves on either side

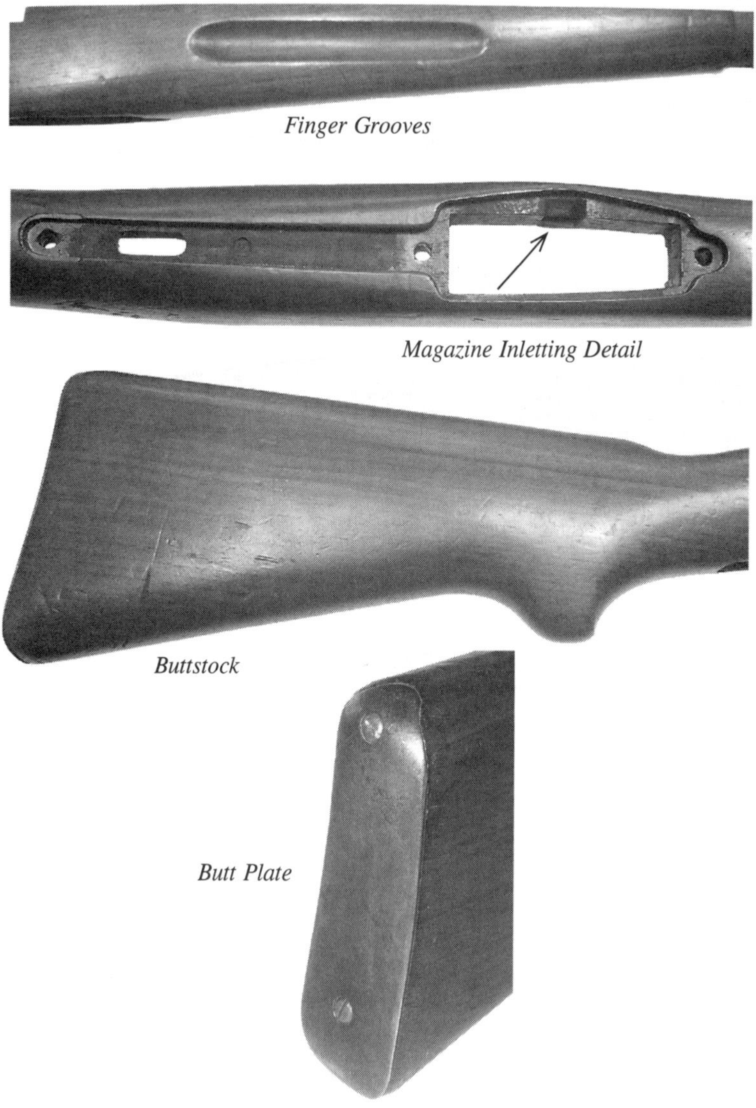

Fig. 3-3. Model 1911 Schmidt rifle stock detail. Compare to Figure 3-1.

of the stock were five inches long, but varied in length from stock to stock by as much as 0.25 inch.

A hole varying between 0.5 and 0.65 inch in diameter was drilled in the end of the buttstock for the stock shaper spindle tooling.

In all other respects, the stock assemblies for the Model 1889 and 1911 series are the same.

Stock Assembly, Model 1911 Carbine

The carbine stock assembly differed from the Model 1911 rifle stock assembly in three particulars. 1) Its overall length was 40.9 inches rather than 49 inches and 2) the rear sling swivel position was changed for the convenience of mounted troops and was inset into the left side of the buttstock, see Figure 3-4. 3) The magazine well inletting resembled that cut into the Model 1911 stock except that inletting for the magazine release was open at the top, see Figure 3-5. Model 1911 carbine stocks were made of European walnut.

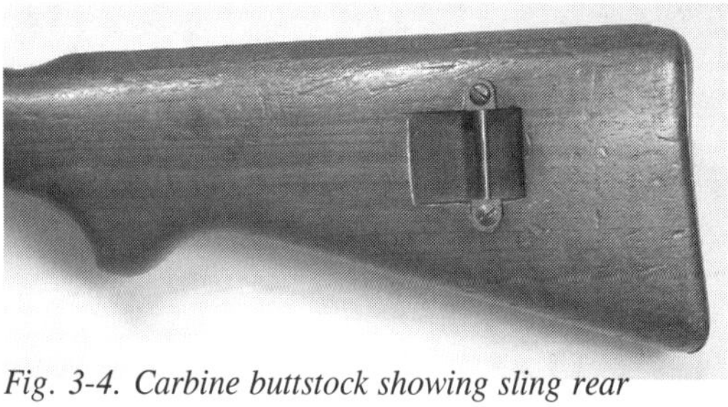

Fig. 3-4. Carbine buttstock showing sling rear attachment point.

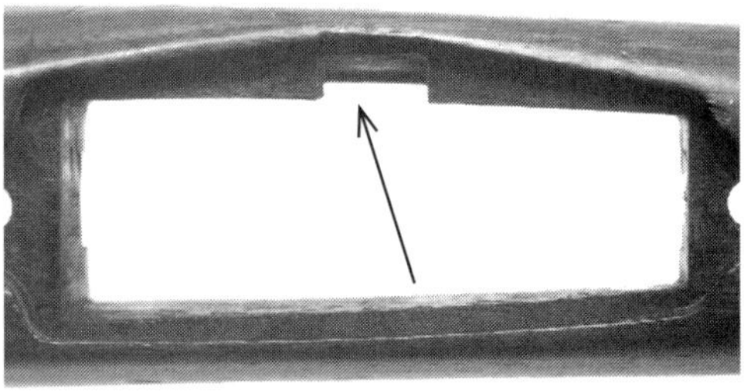

Fig. 3-5. Magazine inletting in the Model 1911 stock.

Stock Assembly, K31

The K31 stock was 41.5 inches long from nose to toe, see Figure 3-6. It was a modifica-

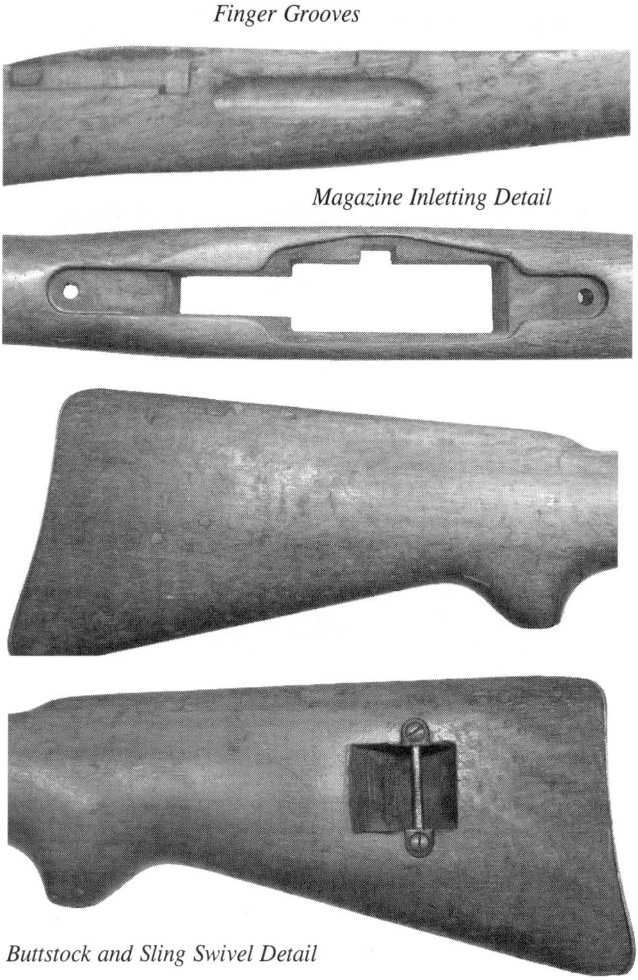

Fig. 3-6. K31 stock detail. Compare to Figures 3-1 and 3-3.

tion of the Model 1911 carbine stock. The length of the trigger guard inletting was 8.31 inches and the shape of the milling for the trigger guard and magazine well was quite a bit different. The adoption of a shorter bolt made it possible to shorten both the receiver and move the magazine well to the rear.

The rear sling swivel was mounted on the left side of the buttstock, as it had been on the Model 1911 carbine buttstock. The finger grooves were offset on either side of the stock. The right finger groove was 3.5 inches long and began 20 inches behind the nose. The left finger groove was also 3.5 inches long but began 17.3 inches behind the nose. The placement allowed gloved or mittened fingers to fall naturally into the finger grooves.

The K31 barrel did not use a barrel collar to free-float the barrel and so there were no recesses in either the stock nose or the front end of the hand guard for it. Very early K31 stocks were made of European walnut but the majority were made of beech, which can be identified by the lighter-colored, almost vertical flecks scattered throughout the grain, and by its yellowish to brown color when oiled. Examine the grain carefully as some of the walnut used for stocks was sapwood, which takes on a yellowish color when oiled. Walnut grain is tight and parallel without the vertical flecks.

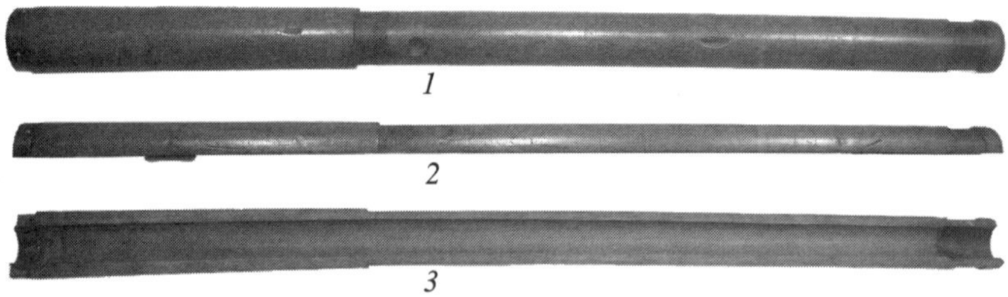

Fig. 3-7. Model 1889 hand guard: 1) top view, 2) side view, 3) bottom view.

Hand Guard

Model 1889 Rifle

Fig. 3-8. Model 1889 barrel and hand guard: 1) locating lug, 2) rear ferrule lip.

The M1889 hand guard, also used on the Model 1889/96, was a one-piece hand guard that extended to within 2.75 inches of the muzzle but did not cover the receiver ring at the rear, see Figure 3-7. The hand guard was positioned on the rifle barrel by a lug on the top of the barrel forward of the rear sight, which fitted into a detent on the inside of the hand guard, see Figure 3-8. The rear and front barrel bands secured the hand guard at those positions. The rear of the hand guard was reinforced with a metal ferrule. A lip on the ferrule slid under the front of the rear sight to hold it in place. The ferrule was riveted to the hand guard with two small steel rivets. The ferrule was blued. The front of the hand guard on the inside was relieved for the barrel collar that free-floated the barrel, see Figure 3-9.

Fig. 3-9. Model 1889 hand guard and stock nose. Note the inletting for the barrel collar.

Hand Guard, Model 1911 Rifle

The Model 1911 rifle hand guard was nearly identical to that used on the Model 1889 rifle. The Model 1911 rifle barrel did not have the projecting barrel lug ahead of the rear sight that positioned the hand guard in the Model 1889 rifle, and so the Model 1911 hand guard lacked the inletted slot on the underside, rear of the hand guard.

Hand Guard, Model 1911 Carbine

The Model 1911 carbine hand guard was lengthened (19.75 inches long) to extend back around the rear sight to cover the front of the receiver ring. A spring clamp at the rear snapped around the barrel. The spring clamp was fastened to the hand guard with two brass rivets, see Figure 3-10.

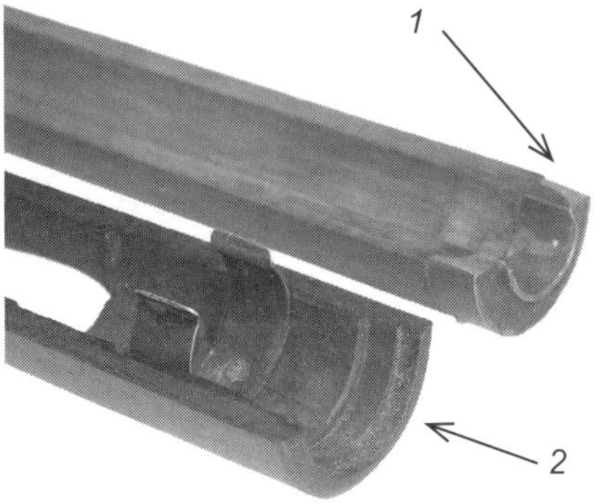

Fig. 3-10. Model 1911 carbine hand guard: 1) inletting for barrel collar, 2) spring clamp at receiver end.

NOTE: To remove the Model 1911 carbine hand guard without splitting the wood, raise the sight leaf. Remove the barreled action from the stock. Turn the barreled action upside down and using your thumbs, press on the ends of the spring clamp to push them off the barrel. Then, turn the hand guard at right angle to the barrel and slide up and over the sight leaf. If you just pull the hand guard up and off, you will probably crack it.

Fig. 3-11. K31 hand guard. Like the Model 1911 rifle hand guard, it does not surround the rear sight and the front of the receiver ring.

Hand Guard, K31 Rifle

The hand guard used on the K31 rifle resembled that used on the Model 1911 rifle but was shorter at 18 inches, see Figure 3-11. Instead of the barrel clamping springs used on the Model 1911 carbine hand guard, the K31 used the Model 1911 rifle metal ferrule with a lip that slid under the front of the rear sight. The K31 hand guard also did not have the inletted recess at the front for the barrel collar.

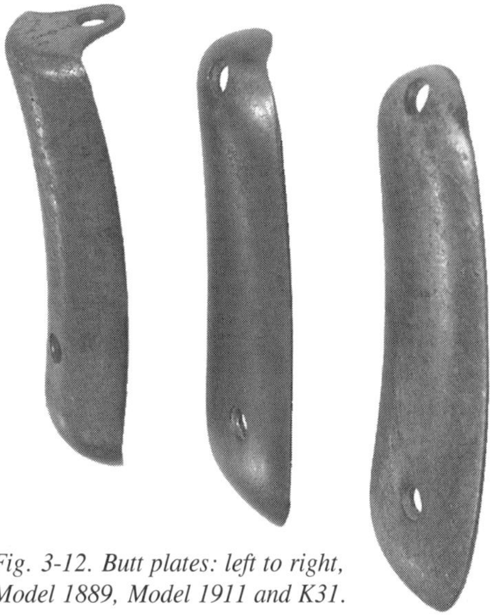

Fig. 3-12. Butt plates: left to right, Model 1889, Model 1911 and K31.

Butt Plate, Model 1889 Rifle

The butt plate used on the Model 1889 rifle was fitted as a cap. The end of the buttstock was relieved to allow the sides of the butt plate to fit over the end. The plate was cast from steel and machined to shape. It was 4.745 inches long and 1.541 inches wide. It had an upper tang 1.165 inches long. The tang was drilled for the upper tang screw. The lower tang screw hole was drilled 1.50 inches above the toe. Butt plates for the Models of 1889, 1911 and the K31 are shown in Figure 3-12.

Fig. 3-13. Butt plate for Model 1911 rifle and carbine.

Butt Plate, Model 1911 Rifle

The Model 1911 butt plate was reduced in size and weight and was no longer fitted as a cap. The prominent tang of the Model 1889 butt plate gave way to a rudimentary tang, and the upper tang screw hole was moved to where the tang curved into the vertical section of the butt plate, see Figure 3-13.

Butt Plate, Model 1911 Carbine

The Model 1911 carbine used the same butt plate as the Model 1911 rifle.

Butt Plate, K31

The K31 butt plate was modified from the Model 1911 butt plate. As the stock butt had a deeper curve than previous models, the butt plate followed that curve. The rudimentary tang at the heel was also eliminated, see Figure 3-14.

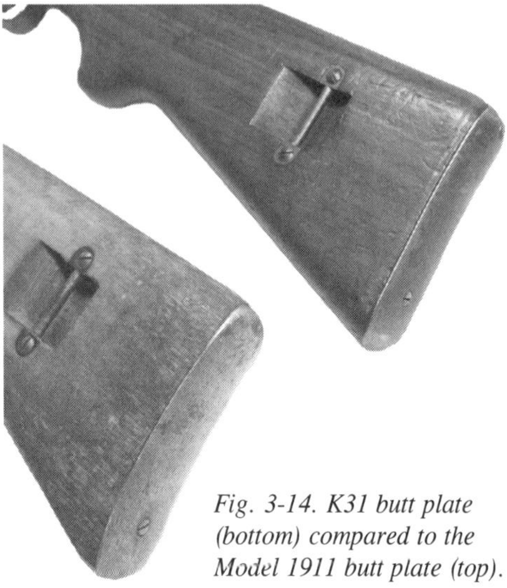

Fig. 3-14. K31 butt plate (bottom) compared to the Model 1911 butt plate (top).

NOTE: Collectors and shooters owning a K31 rifle might want to remove the butt plate to check for a slip of paper or cardboard with the name and address of the Swiss soldier to whom it was issued. As Swiss soldiers usually kept their rifles and other military gear at home, they often added the slip of paper with their name and address so that the rifle could be returned to them if misplaced, lost or stolen. See Figure 3-15.

Fig. 3-15. Name and address of Swiss soldier to whom this K31 rifle was issued.

Barrel Bands, Model 1889 Rifle

Two barrel bands in addition to the four guard screws fastened the stock to the receiver, see Figure 3-16. The forward barrel band was 1.819 inches high and 0.7 inch wide. It was hinged on

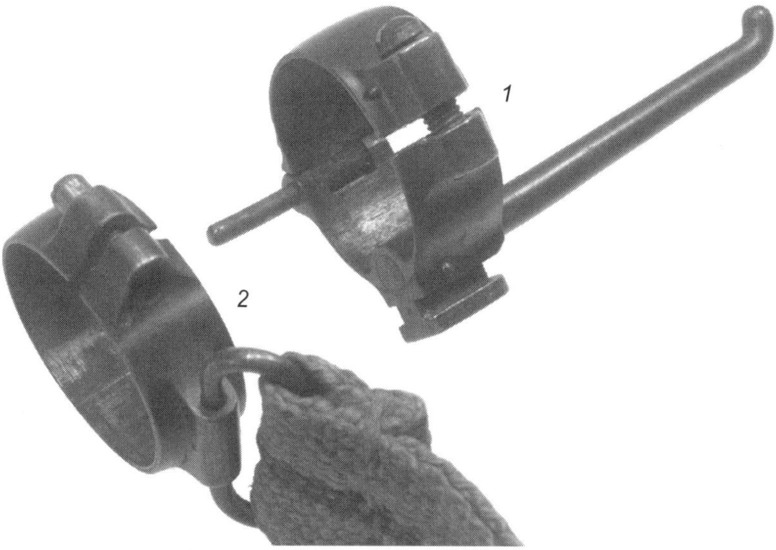

Fig. 3-16. Model 1889 upper (1) and (2) lower barrel bands.

the left side and held together on the right side with a vertical screw through its upper and lower lugs.

The bayonet mount was machined from the lower part of the band. The stacking rod was fitted into a socket on the inside of the bottom part of the band and held in place with a pin, just above the Mauser-style bayonet mount. The stacking rod curled up slightly.

The rear band was a clamping barrel band with a lug on either side of the split right side. The lugs were drilled and threaded for the barrel band screw. This screw had a threaded shank and a round collar 0.055 inch thick and 0.256 inch in diameter. The upper lug was drilled to accept that collar which, when tightened down, bottomed against the inside of the lug. The top half of the lug was 0.395 inch thick and the bottom half was 0.270 inch thick. The screw had a short shank above the collar and a slotted head. The clamping screw was held in the top clamping lug by a pin that passed through the lug and between the screw's collar and slotted head.

The forward sling swivel was mounted on the bottom of the lower barrel band, was 1.66 inches long and formed from iron wire 0.136 inch in diameter. The lower band was held in place against recoil with a barrel band spring inletted into the forend ahead of the band. The barrel band spring was fastened to the stock with a screw that penetrated through the barrel channel.

Barrel Bands, Model 1911 Rifle

The Model 1911 rifle barrel bands were identical to those used on the Model 1889 rifle.

Barrel Bands, Model 1911 Carbine

The front barrel band used on the Model 1911 carbine was identical to that used on the Model 1911 rifle but the lower band differed, see Figure 3-17, in that the sling swivel was mounted on the left side. The carbine lower band was split at the bottom and a clamping screw, similar to that used on the rifle lower band, passed through two lugs milled on either side of the split. A lug on the left side of the lower band was drilled from top to bottom for the sling swivel. The sling swivel was slightly smaller at 1.57 inches long. The diameter of the wire was the same. The lower band was held by a band spring (see arrow).

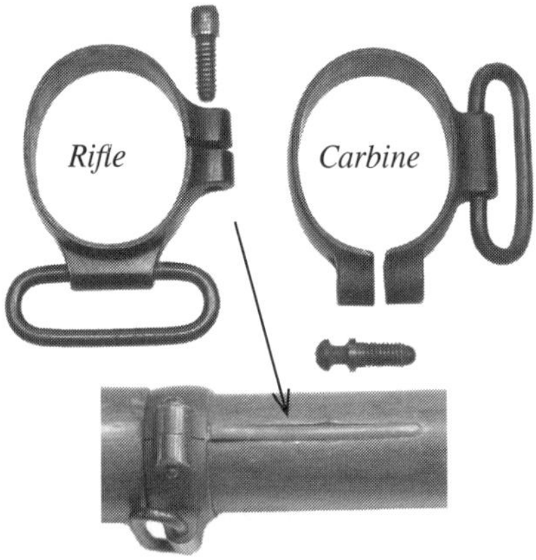

Fig. 3-17. Model 1911 rear carbine barrel band compared to Model 1911 rear rifle barrel band.

Barrel Bands, K31

The barrel bands used on the K31 were similar to those used on the Model 1911 carbine. The front band had a smaller clamping screw and the lugs were 0.283 inch thick. The K31 and Model 1911 carbine lower barrel bands were identical.

Trigger Guard Assembly, Model 1889 Rifle

While there are common features in the trigger guard assemblies for the three major models, Model 1889, Model 1911 and K31, they did change as the rifle was reduced in size and complexity. See Figure 3-18, top view and Figure 3-19, bottom view of the three basic trigger guards.

The Model 1889 trigger guard plate was 10.5 inches long and was held to the receiver with three guard screws, two short (a) and one long (b) which passed through the stock into the receiver, and one wood screw (c) which fastened the back of the trigger plate to the stock. All three trigger guard plates housed the magazine.

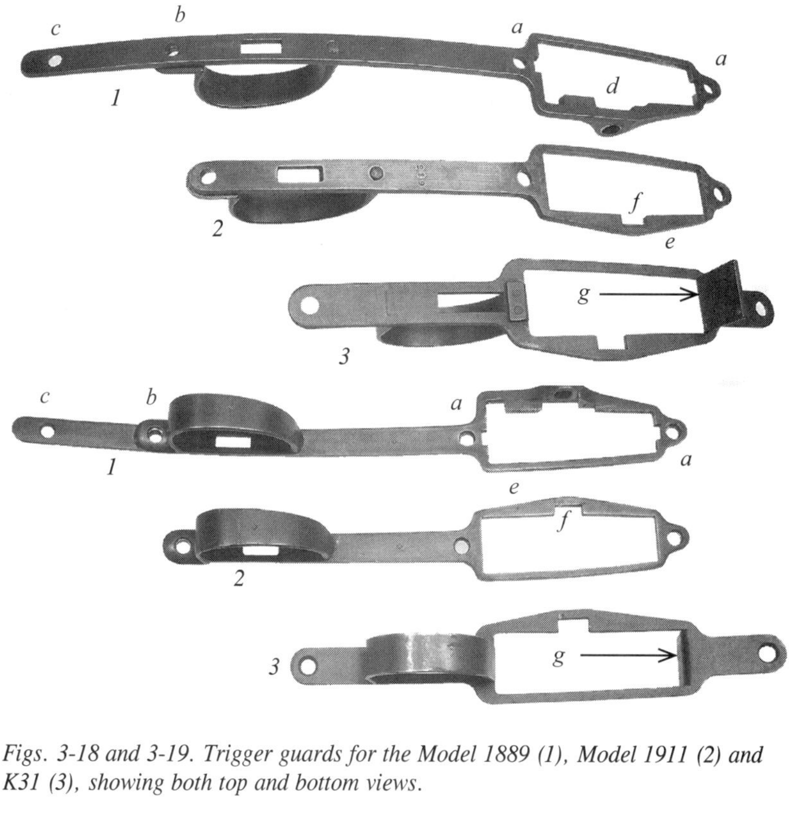

Figs. 3-18 and 3-19. Trigger guards for the Model 1889 (1), Model 1911 (2) and K31 (3), showing both top and bottom views.

The Model 1889 trigger guard plate and trigger guard were made separately. The front post of the trigger guard was threaded and screwed into the trigger guard plate. The rear of the trigger guard was fastened to the plate with the long guard screw.

A lug (d) was shaped on the right side of the trigger guard plate for the magazine cutoff switch mechanism, see Figure 3-20. The lug was 0.44 inch high with a 0.311-inch-diameter hole for the magazine cutoff switch pivot screw.

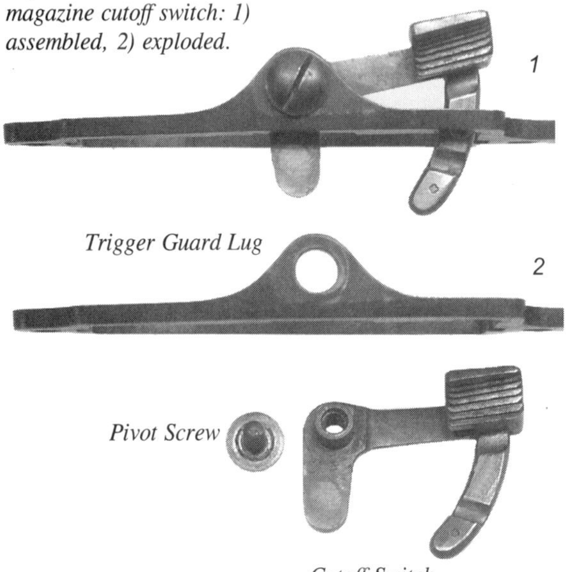

Fig. 3-20. Model 1889 magazine cutoff switch: 1) assembled, 2) exploded.

Trigger Guard Assembly, Model 1911 Rifle and Carbine

The Model 1911 trigger guard plate differed from the Model 1889 trigger guard plate in three particulars, refer to Figures 3-18 and 3-19, top and bottom views. First, it was reduced in length from 10.5 to 9.6 inches long. Secondly, the lug for the magazine cutoff was eliminated. Instead, the right side of the magazine well was bowed outward (e) and a rectangular cut was made for the magazine release switch (f). Third, the fourth wood screw used in the Model 1889 was eliminated. The Model 1911 trigger plate assembly retained the separate trigger guard.

Trigger Guard Assembly, K31

The K31 trigger guard was shortened considerably with the adoption of the new bolt and receiver which moved the magazine to the rear, hard against the trigger guard, refer to Figures 3-18 and 3-19, top and bottom views. The K31 trigger guard was 8.75 inches long and stamped from sheet steel. The trigger guard was welded permanently to the trigger guard plate. A flap of steel sheet 0.94 inch high and as wide as the magazine opening at the front (0.94 inch wide) was turned up vertically to form the front magazine guide (g). Only two stock screws were used to hold the trigger guard plate in place.

Magazine, Models of 1889 and 1889/96

At a time when many European nations were selecting the five- or six-round Mannlicher magazine system, the Swiss adopted a twelve-round box magazine, based on the Lee patents, that could be fully loaded with two chargers or clips of six cartridges each, or single loaded one cartridge at a time, see Figure 3-21. The Model 1889 box magazine entered a cutout in the trigger guard plate and was not removable unless the entire trigger guard assembly and magazine cutoff switch were removed.

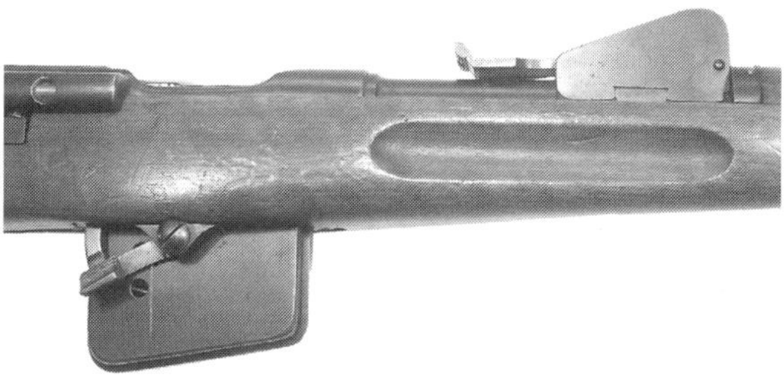

Fig. 3-21. Model 1889 rifle showing the 12-round box magazine and cutoff switch. The switch is down, which raises the magazine to feed cartridges.

The twelve-round box magazine was formed from sheet steel and held twelve cartridges. A reinforcing steel strap was brazed to the magazine from front to back. The magazine also had three holes on either side to allow dirt and moisture to escape, and to help the soldier keep track of the number of cartridges remaining. A wire Z-spring was used to force the cartridge guide up against the lips of the magazine that were folded over from the sides, see Figure 3-22. Cartridges were loaded through the open breech by inserting a cartridge clip made of pressed paper and metal into the clip guide at the front of the receiver bridge. The M1889 and its succeeding variations did not have a charger guide as such; rather, the opening in the top of the breech for the cartridges was shaped to hold the entire charger.

The parts of the magazine assembly were: 1) trigger guard plate, 2) trigger guard, 3) follower, 4) magazine spring, 5) magazine box, 6) cutoff switch, 7) cutoff switch pivot screw, 8) reinforcing strap and 9) drainage holes.

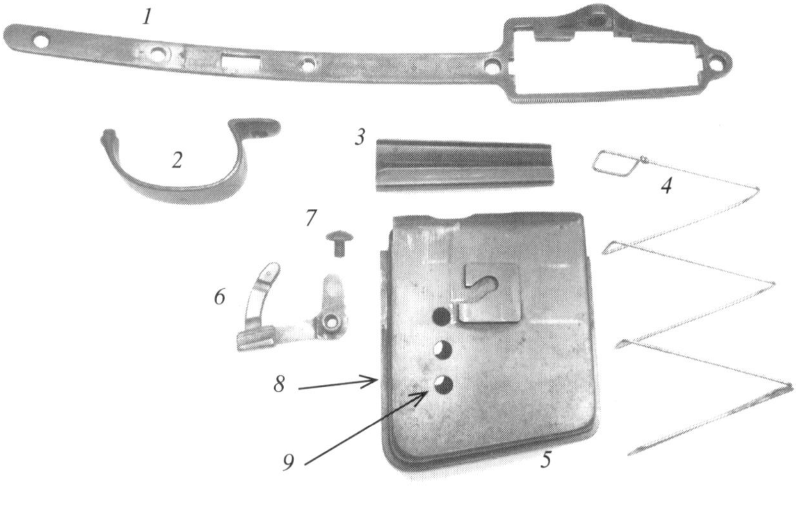

Fig. 3-22. Parts of the Model 1889 rifle magazine.

The magazine follower was a curved steel platform with the right side higher than the left to force the cartridges into a staggered column in the magazine. The cartridges were pressed down with the thumb past the magazine lips into the magazine. The process was repeated with a second clip to fill the magazine to capacity. When the bolt was pushed forward, the bolt face pushed the top cartridge into the breech.

Most magazine cutoff systems of the time prevented a cartridge from being moved from the magazine into the breech either by limiting the bolt travel or by covering the magazine lips. The Schmidt system actually lowered the magazine so that the top cartridge in the magazine was below the line of bolt travel, see Figure 3-23. A lug on the lever system attached to the left side of the trigger guard assembly engaged a detent in the right side of the bolt body and raised or lowered it when the magazine switch was moved.

Fig. 3-23. When the cutoff switch was up (as shown), the magazine was moved down below the path of the bolt face and a cartridge did not feed.

Magazine, Model 1889/1900 and Model 1905 Carbine

This model introduced a new detachable box magazine reduced in capacity to six cartridges and did away with the magazine cutoff system, see Figure 3-24.

The parts of the magazine were: 1) reinforcing ridge, 2) reinforcing strap and 3) magazine release switch.

The new magazine was quite similar to the box magazine developed by James Lee and used on the British Lee-Enfield rifle. The magazine was made of stamped steel with two reinforcing ridges crimped into the bottom of either side. The bottom of the magazine was flat and angled upward toward the front. A wire Z-spring moved the magazine follower up against the feed guides formed from the

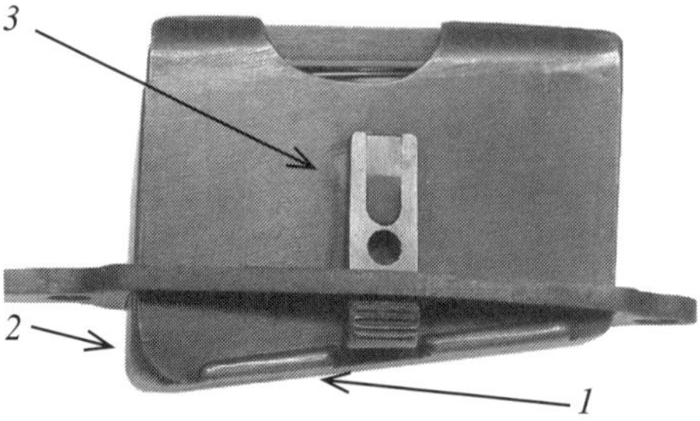

Fig. 3-24. Model 1889/1900 Lee-type box magazine.

sides of the magazine. The follower was similar to that used in the Model 1889 and forced the six cartridges into two staggered columns. The magazine was held in the magazine well by a rocker switch on the right side of the magazine. The switch is a straight piece of steel, rounded at the end into a thumbpiece. Directly above the thumbpiece is a cut that snaps into a detent in the trigger guard plate. The switch is forced outward by a small coil spring pressing against the magazine body.

The Model 1889/1900 magazine had a reinforcing steel strap running from front to rear as did the Model 1889 twelve-round magazine. It also had four small holes in the bottom to allow dirt and moisture to drain out. Later magazines eliminated the steel reinforcing strap and the drain holes.

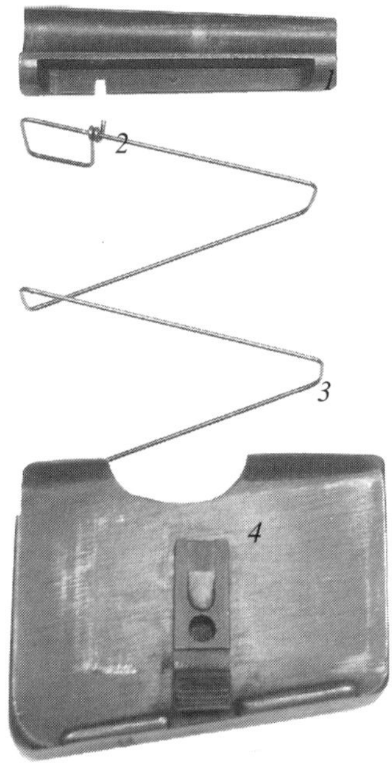

Fig. 3-25. Model 1911 magazine.

Model 1911 Magazine

This magazine differed from the Model 1889/1900 magazine only in that it lacked the reinforcing steel strap and the drain holes, see Figure 3-25.

The parts of the Model 1911 magazine were: 1) follower, 2) magazine spring, 3) magazine body and 4) magazine release switch.

K31 Magazine

This magazine differs quite a bit in shape from the Model 1911 magazine and they are not interchangeable. The K31 magazine is made from stamped steel with a rounded bottom that angles slightly upward toward the front, see Figure 3-26. The magazine switch, while still mounted on the right side, is set back farther toward the rear than the Model 1911 magazine and rather than attaching to the trigger guard plate, its locking groove attaches to a detent in the receiver for a more positive hold. The magazine box has a vertical reinforcing crimp on either side just ahead of the magazine switch.

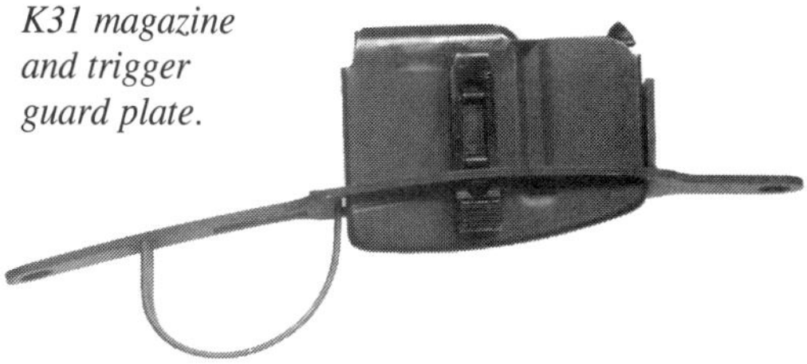

Fig. 3-26. The K31 magazine and trigger guard plate.

The cartridge guide is also shaped differently. In cross section, it has an "S" shape for half its length. The front was pressed into a raised semicircle.

The Z-spring was changed from wire to flat spring steel and was riveted to the bottom of the cartridge

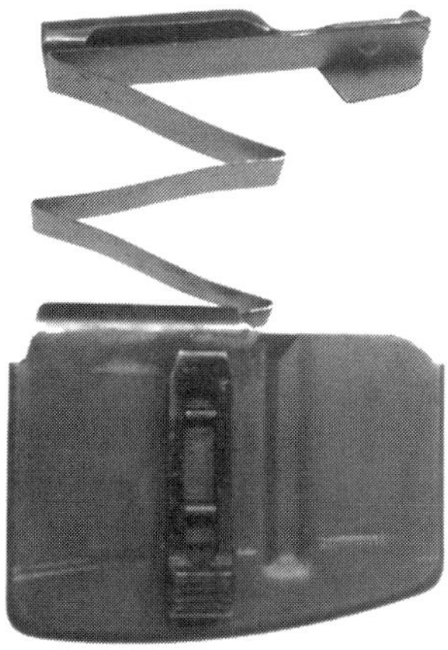

Fig. 3-27. A flat magazine spring replaced the wire spring in the K31 magazine.

Receiver, Model 1889 Rifle

The receiver was forged from a steel bar and machined to shape, see Figure 3-28. It was a complicated structure, which included the tube (a) for the bolt ran as well as an auxiliary tube (b) on the right in which the bolt handle functioned. Two lightening cuts (c) were milled into the top rear behind the opening in the top of the receiver, as well as a single lightening cut (d) on the left side of the receiver. The magazine well (e) was cut into the receiver. It was rectangular with the front end rounded. Two square cuts were made at the front and rear of the magazine well to accommodate the reinforcing band around the periphery of the magazine (f). The back of the receiver extended into the receiver tang (g) in which the trigger was pinned.

The Model 1889 receiver was 9.3 inches long. The ejector was a round pin set (h) into the left side of the receiver wall at the front edge of the bridge. It was secured in placc with a sctscrcw.

The trigger (i) was a simple three-part assembly, see Figure 3-29. The trigger (1) was attached to the sear lever (2) with a pin (3). The sear (4) protruded from the sear lever and through a slot (5) in the bottom of the receiver tang (6). When the bolt was cocked, the cocking lug (7) was held back against the main spring by the sear. When the trigger was pulled, the bearing point (8) on the top of the trigger lever bore against the bottom of the receiver (9). This caused the sear lever to rotate

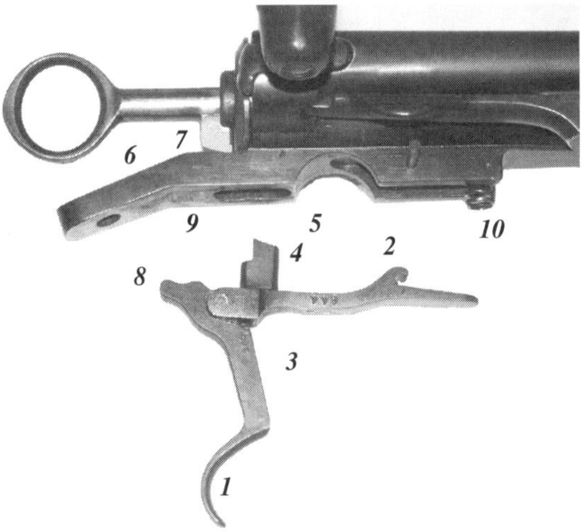

Fig. 3-29. Schmidt trigger/sear assembly.

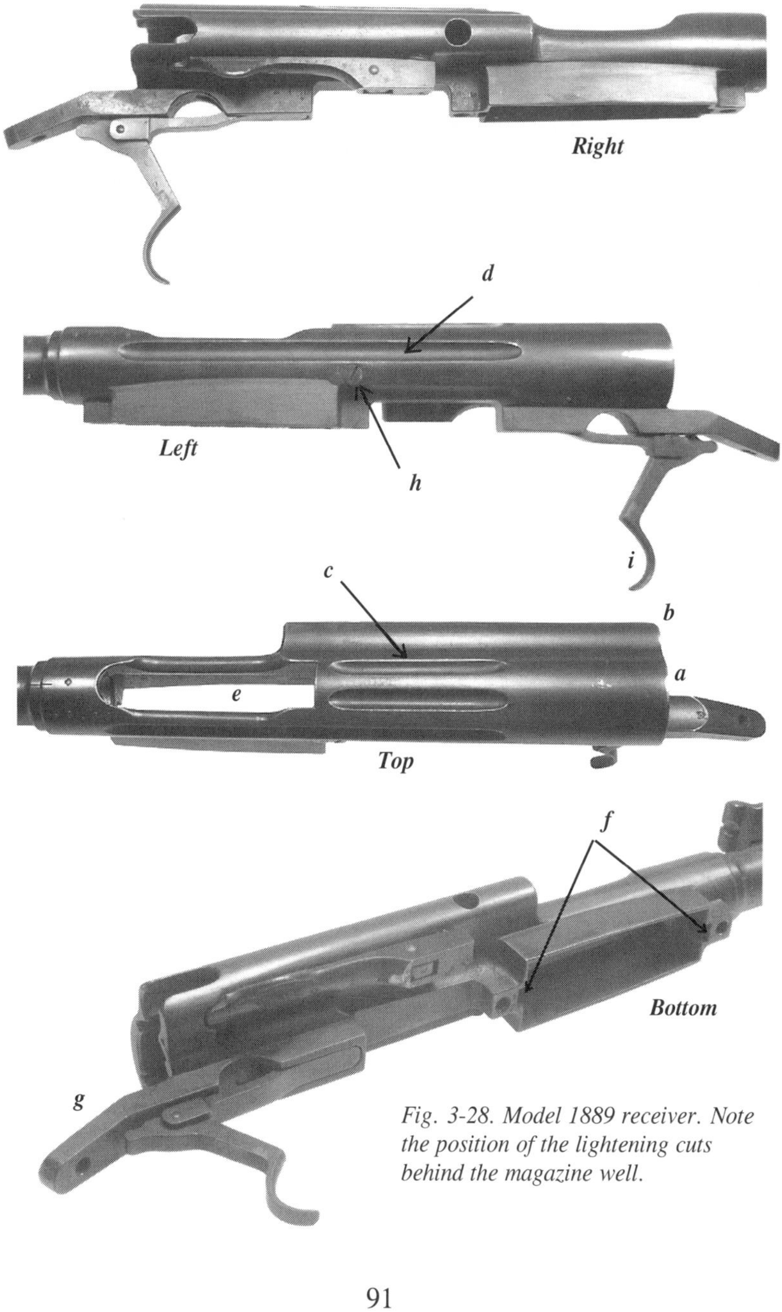

Fig. 3-28. Model 1889 receiver. Note the position of the lightening cuts behind the magazine well.

down on its axis point, pulling the sear out of engagement with the cocking lug. The coil spring (10) between the end of the sear lever and the receiver caused the sear lever to rotate back up, moving the sear up into its receiver slot, ready to cock the bolt again.

The Model 1889 receiver was modified slightly for the Models of 1889/96, Model 1889/00 and Model 1905 to move the recesses for the bolt locking lugs ahead in the receiver to just behind the ejection opening.

Receiver, Model 1889/1896, Model 1889/1900, Model 1905, and Model 1911 Rifle and Carbine

The receivers used on the Schmidt-design rifles, following the Model 1889, are very much alike with only two minor variations in shape. The new receiver was reduced in length to 8.46 inches, see Figure 3-30. Moving the bolt locking lugs to the front of the locking cylinder allowed the receiver to be shortened by 0.79 inch. In addition, it was strengthened with the addition of metal in areas of stress. The relief cuts (a) at the front and rear of the magazine well for the magazine reinforcing band were dispensed with in the Model 1911 receiver as its magazine did not have a reinforcing band. Model 1911 lightening grooves (b) were moved to the rear of the receiver. In all other respects, the receiver remained the same for the Models of 1889/96, 1889/00, Model 1905 carbine and Model 1911, except for markings.

Receiver, K31

The K31 receiver was shortened another 1.9 inches when the bolt assembly was redesigned. The bolt handle tunnel was flatter and lacked the round hole found in the bolt handle tunnel of the Model 1911 receiver. The bolt release lever was stamped from sheet steel rather than milled from a steel forging. The top, rear of the receiver was flatter and there were no lightening grooves milled into the metal. The slot for the sear lug was moved ahead 0.36 inch and the magazine well was moved rearward, see Figure 3-31.

A new trigger assembly was designed for the shorter receiver, see Figure 3-32. It was housed in an open box and the sear became a vertical lever which pivoted at a point two-thirds of its length. The

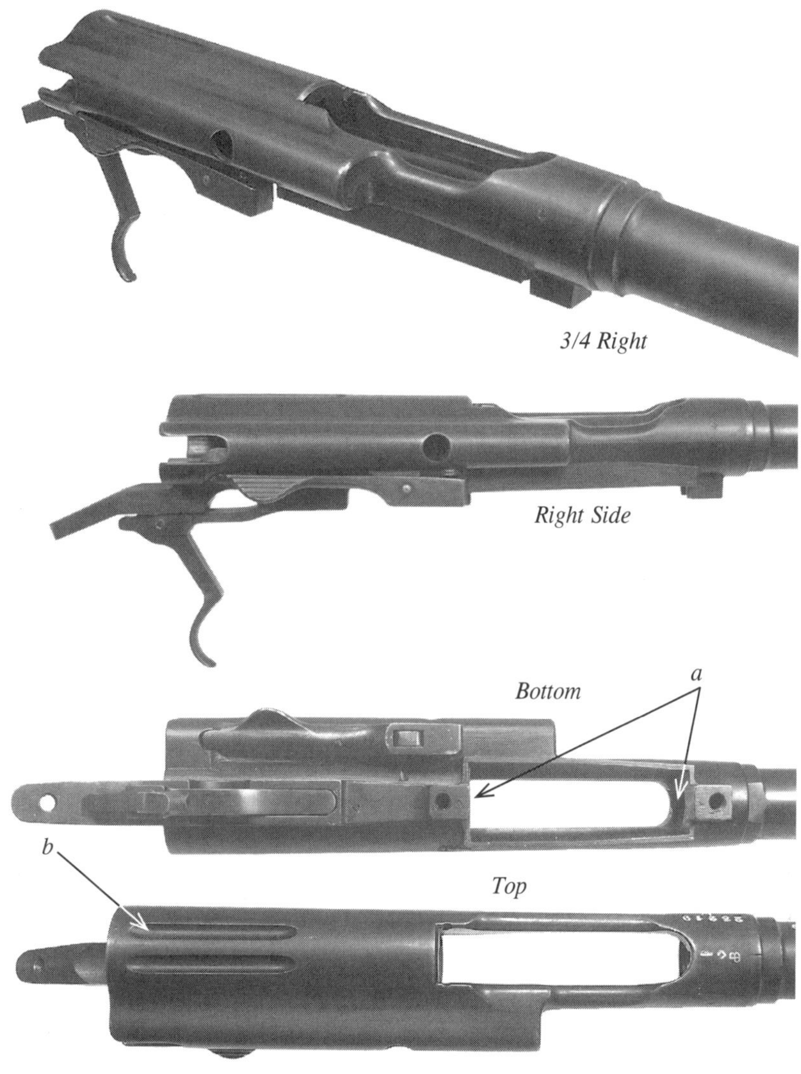

Fig. 3-30. Model 1911 receiver. Note the position of the lightening cuts at the rear of the receiver.

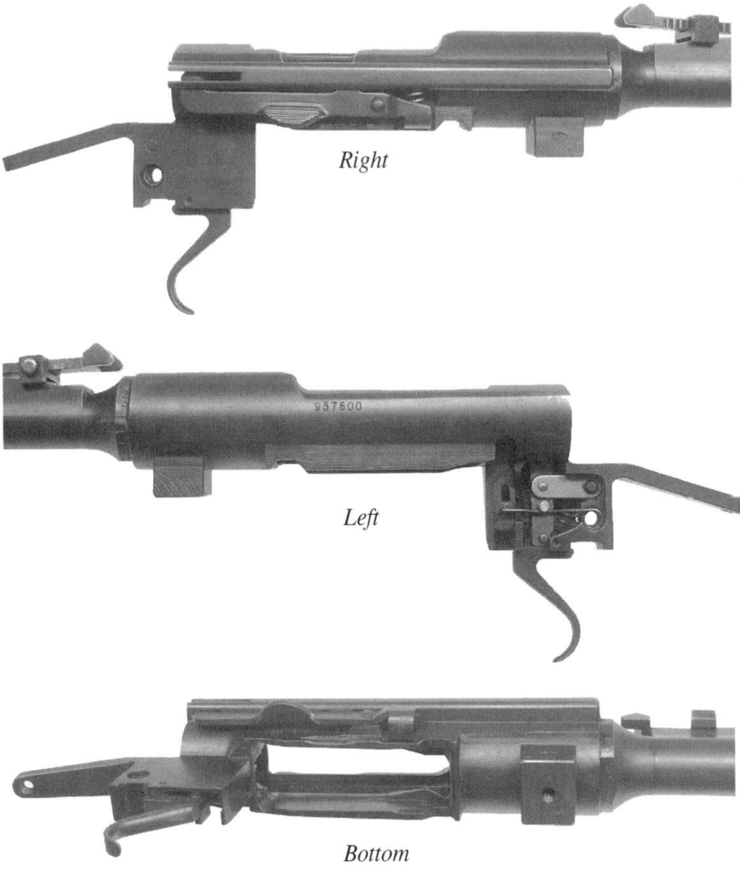

Fig. 3-31. K31 receiver. Note the lack of lightening cuts.

bottom of the sear was pinned to the to the top of the trigger. When the trigger was at rest, the trigger/sear spring pushed the sear up through its slot in the receiver to make contact with the cocking lug on the striker and hold it back against the main spring. When the trigger was pulled, it rotated down, pulling the sear out of contact with the striker and allowing it to be pushed forward by the main spring. The two-stage trigger design operated smoothly, and the break between the first and second stage was positive.

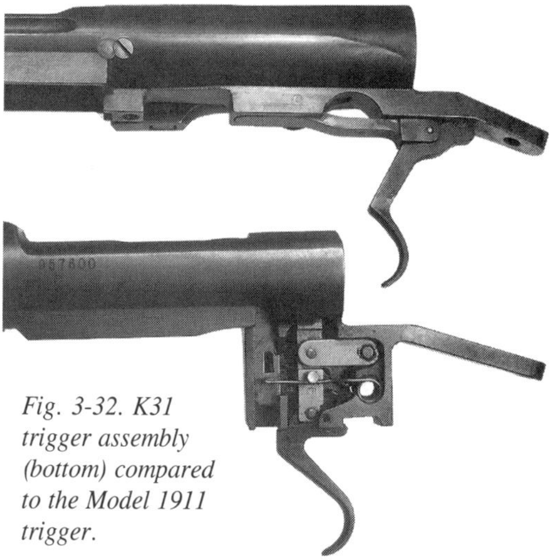

Fig. 3-32. K31 trigger assembly (bottom) compared to the Model 1911 trigger.

Bolt Assembly, Model 1889

The most unique aspect of the Schmidt rifle was the non-rotating bolt assembly designed by Colonel Schmidt, which combined the Mannlicher straight pull bolt system with a cam-actuated version of the Vetterli bolt mechanism, see Figure 3-33. The original four-part bolt in the Model 1889 was 11 inches long. Figure 3-34 shows the bolt assembly exploded: 1) bolt cylinder, 2) locking sleeve with the locking lugs, 3) bolt cap, 4) striker, 5) cocking lug, 6) firing pin, 7) main spring, 8) extractor and 9) bolt handle.

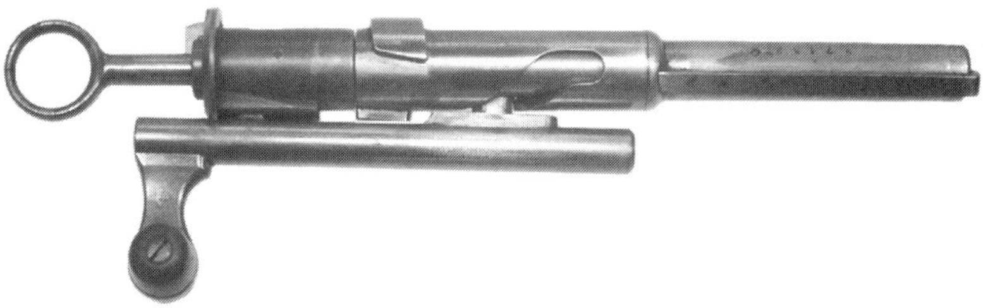

Fig. 3-33. Model 1889 bolt assembly. Note the lugs mounted well to the rear of the bolt's center point.

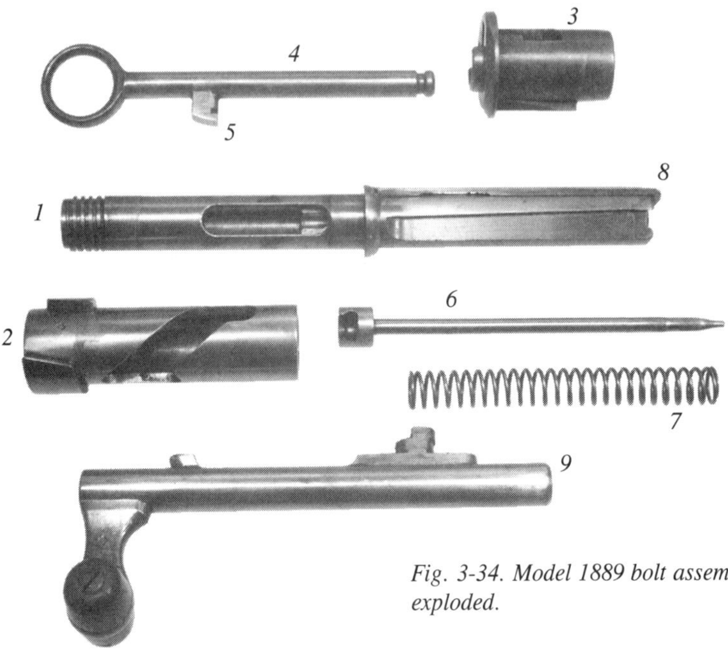

Fig. 3-34. Model 1889 bolt assembly exploded.

The bolt cylinder was a combination of a hollow tube and long forward bar with a separate bolt face permanently attached, see Figure 3-35. The bolt cylinder was bored straight through for the firing pin and spring. The cylindrical portion to the rear was large enough in

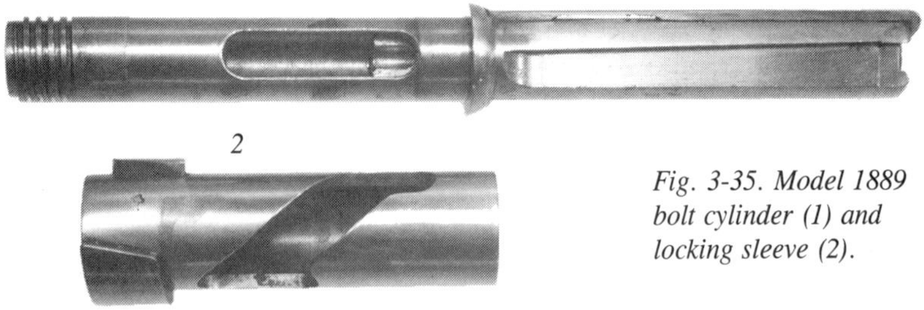

Fig. 3-35. Model 1889 bolt cylinder (1) and locking sleeve (2).

diameter for the striker collar to pass. A rectangular cut with rounded ends 1.5 inches long was cut into the side for the lug on the bolt cam follower rod. The firing pin, which was 4.75 inches long, was joined to the 5.5-inch-long cocking piece by a slip joint very much like that in the U.S. Model 1892-1896-1898-1899 Krag and the Model 1903 Springfield.

The main spring was made of round spring steel wire 0.05 inch in diameter. It was 3.8 inches long, 0.461 inch in diameter and had twenty-seven coils.

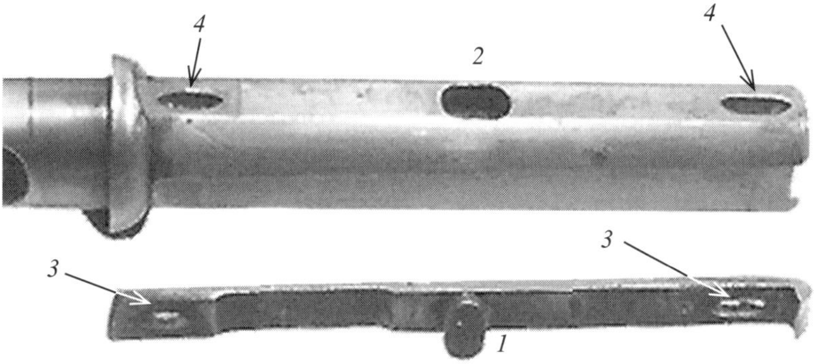

Fig. 3-36. Bolt cylinder and extractor. The extractor lug (1) fits into the lug hole on the bolt cylinder (2). Projections on the bottom of the extractor (3) match detents on the bolt cylinder (4).

The extractor was a flat piece of spring steel 3.6 inches long, see Figure 3-36. Its lip at the front curled over the bolt face to hold the cartridge rim securely. It was held in a dovetail on the left side of the bolt cylinder. Both sides of the bolt cylinder's long forward extension were grooved to permit it to pass through the magazine lips.

The striker had a large finger ring at the back, see Figure 3-37, which served

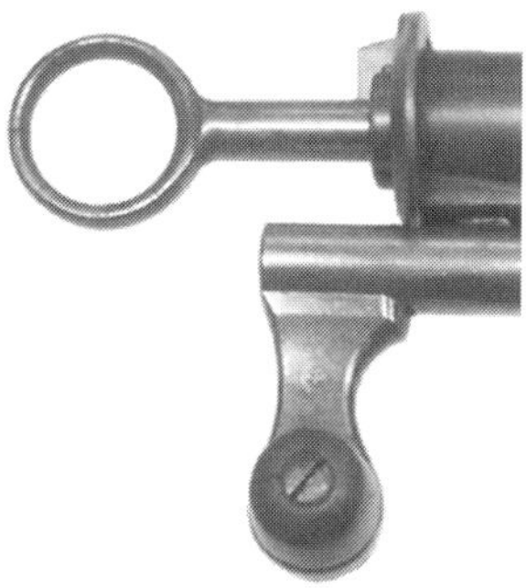

Fig. 3-37. Striker finger ring.

three purposes. 1) It allowed the firing pin to be reset in case of a misfire without having to work the bolt and waste a cartridge. 2) The ring was also the rifle's safety. When pulled back and turned clockwise, 1/4 of a turn, the striker was withdrawn and the cocking lug entered a safety slot in the bolt cap. 3) It also prevented the bolt from being opened.

The ring on the end of the cocking piece protruded 2.25 inches when the rifle was in battery and was a very visible cocking indicator. The cocking lug was mounted on the bottom of the cocking piece and engaged the sear when the rifle was cocked, see Figure 3-38. The track allowed the lug on the operating handle and bolt cam to engage the track in the bolt cylinder.

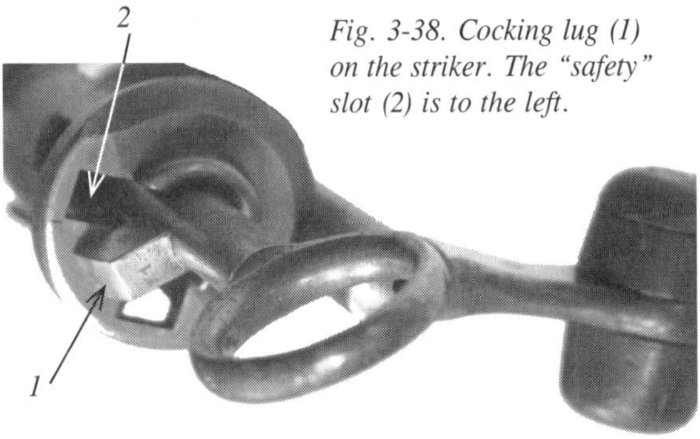

Fig. 3-38. Cocking lug (1) on the striker. The "safety" slot (2) is to the left.

The bolt locking collar, or sleeve, was a steel cylinder bored through to receive the tubular part of the bolt cylinder.

Two large, trapezoidal locking lugs were mounted on the rear, and a spiral track was machined through its side. The front and rear faces of the lugs were cut to follow a track on the inside of the receiver.

The bolt cap held everything in alignment. It was secured to the bolt cylinder by a square thread.

The bolt handle with its cam follower rod traveled in a track in the locking sleeve. The hand piece, or cocking handle, has two vertical knobs to make it easier to grasp. In the Model 1889, the knobs were made of walnut.

Bolt Assembly, Model 1889/1896, Model 1889/1900, Model 1905, Model 1896/1911 and Model 1911

The Model 1911 bolt assembly was shortened to eight inches, see Figure 3-39, and the locking lugs were moved to the front of the locking sleeve to strengthen the assembly for use with the new smokeless cartridge, see Figure 3-40. The bolt face was manufactured as a separate piece and screwed into the front of the bolt body. The vertical knobs on the cocking piece were made of bakelite rather than walnut. In all other respects, the bolt assembly was similar to the Model 1889 bolt assembly.

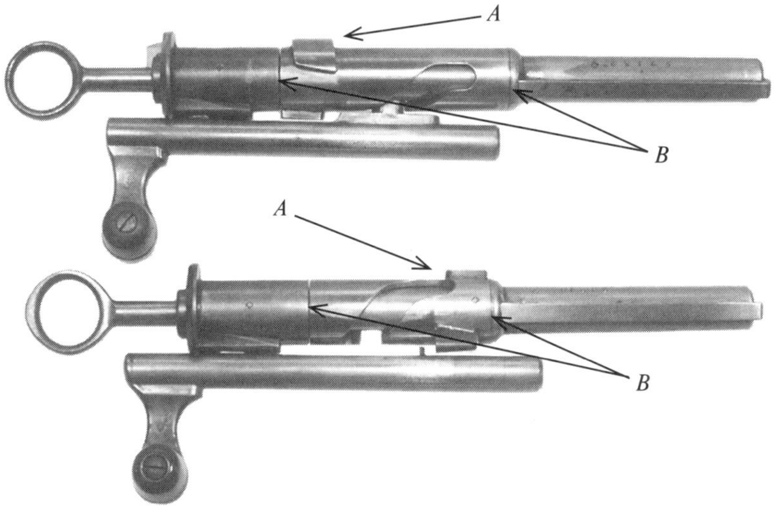

Fig. 3-39. Top: Model 1889 bolt assembly; bottom: Model 1911 bolt assembly. Note the difference in the positioning of the locking lugs (arrow A) and the length of the bolt locking sleeves (arrow B).

The main spring was made of flat spring steel wire and was 2.25 inches long and 0.485 inch in diameter. The wire was 0.04 inch thick by 0.06 inch wide. It had 16 coils.

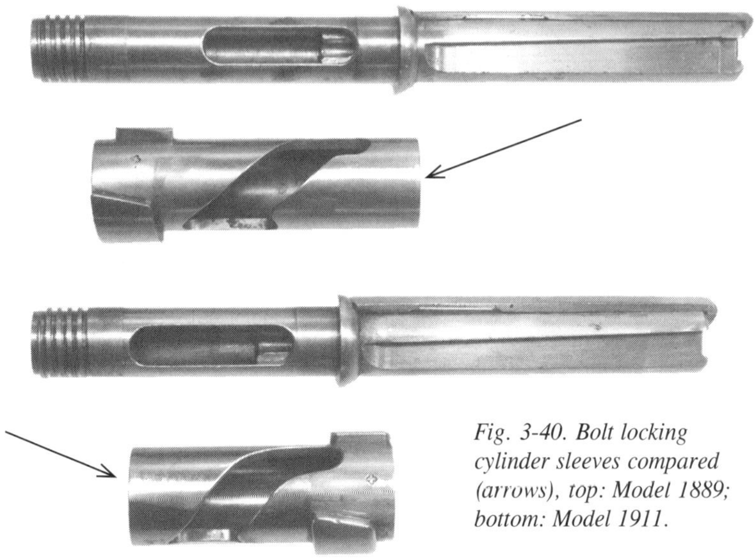

Fig. 3-40. Bolt locking cylinder sleeves compared (arrows), top: Model 1889; bottom: Model 1911.

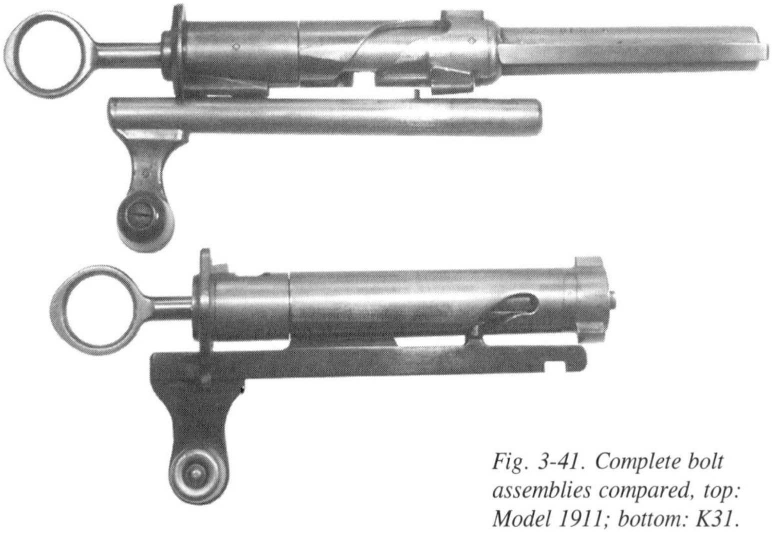

Fig. 3-41. Complete bolt assemblies compared, top: Model 1911; bottom: K31.

Bolt Assembly, K31

The K31 bolt was only slightly more than half as long as the Model 1911 bolt, see Figure 3-41.

The reduction in length was accomplished by shortening the bolt cylinder, and keeping the dual, opposed locking lugs at its front, refer to Figure 3-41 and see Figure 3-42.

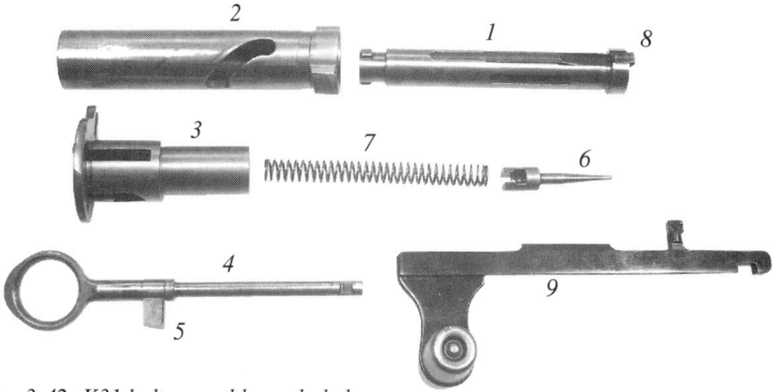

Fig. 3-42. K31 bolt assembly exploded.

The parts are:1) bolt cylinder, 2) locking sleeve, 3) bolt cap, 4) striker, 5) cocking lug, 6) firing pin, 7) main spring, 8) extractor and 9) bolt handle.

The long extension on the bolt body used on the Model 1889 and Model 1911 bolt assemblies was shortened and telescoped into the locking cylinder. The bolt face was no longer applied as a separate piece but was machined into the end of the bolt body. The extractor was shortened and reshaped. It slid into a dovetail in the top of the bolt body and was held in place by a downward projecting stud in a recess behind the bolt face, see Figure 3-43. Aluminum bolt handle knobs replaced the bakelite knobs used previously.

The main spring was 3.3 inches long, 0.335 inch in diameter and made of flat spring steel wire 0.036 inch thick and 0.049 inch wide. It had 34 coils.

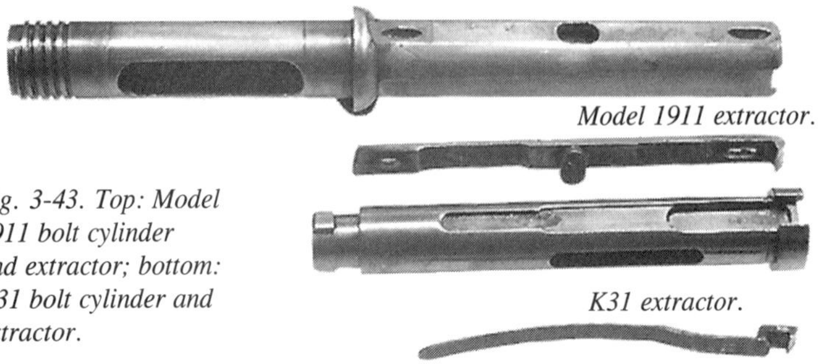

Fig. 3-43. Top: Model 1911 bolt cylinder and extractor; bottom: K31 bolt cylinder and extractor.

Barrel, Model 1889 Rifle

The Model 1889 rifle had a 30.7-inch barrel with two steps. The bore was rifled with one turn in 10.63 inches, see Figure 3-44. The three-groove rifling turned to the right.

The first step was a collar (1) which bore against the receiver face when the barrel was mounted in the receiver. The second step was 2.76 inches long and reinforced the breech (2). On the underside, a flat (3) was cut away 1.38 inches long by 0.055 inch deep. The barrel muzzle was counterbored with a funnel shape to protect the lands and grooves, refer to Figure 1-32. Just behind the front sight, the barrel was swelled slightly to position the barrel collar.

Fig. 3-44. Model 1889 barrel showing double steps between the receiver ring and rear sight.

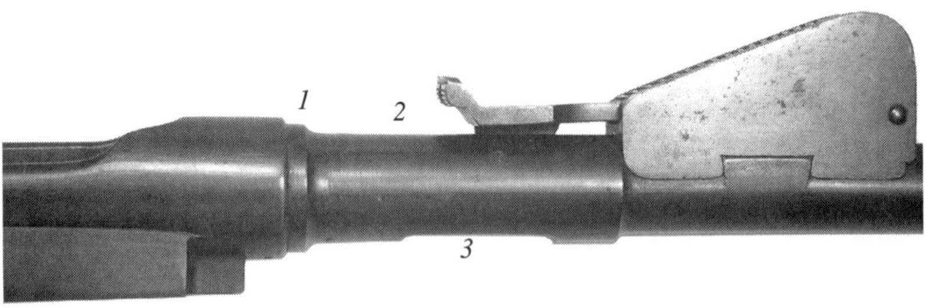

NOTE: All Schmidt rifle barrels of whatever model had a funnel-shaped counterbore.

The rear sight base was mounted on a platform machined from the barrel material. The top of the platform was machined into a dovetail, see Figure 3-45 (arrow 1). The barrel had a straight taper to the muzzle. Immediately in front of the rear sight was a raised lug (arrow 2) on the top of the barrel which positioned the hand guard, also see Figure 3-46 (arrow). The hand guard did not extend behind the rear sight.

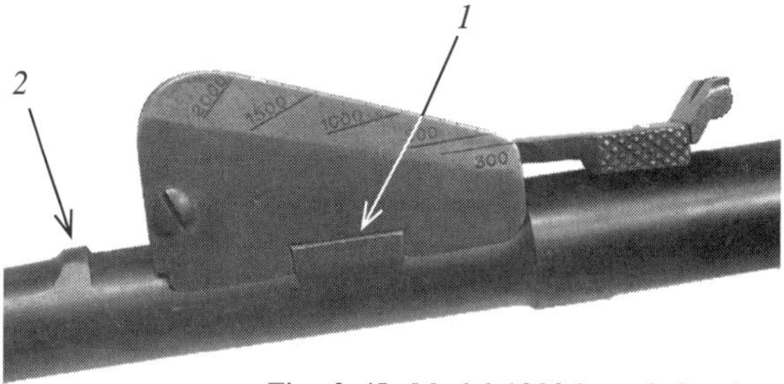

Fig. 3-45. Model 1889 barrel showing rear sight dovetail detail.

The front sight base was a platform machined from a ring which was brazed to the barrel 2.25 inches behind the muzzle, see Figure 3-47. The front sight blade was dovetailed into the base. The rifle was zeroed for windage at the factory by moving the sight blade left or right. Once adjusted, the base of the blade was staked on both sides into the dovetail.

Fig. 3-46. Model 1889 barrel showing lug for positioning the hand guard (arrow).

The Schmidt rifles through the Model 1911 free-floated the barrel with a barrel collar for improved accuracy. The only other military rifle to use a barrel collar to free-float the barrel was the Finnish Model 1928/30

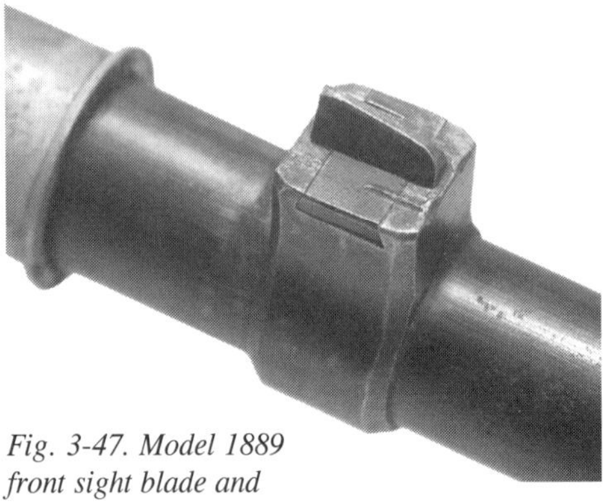

Fig. 3-47. Model 1889 front sight blade and base.

Civil Guard Rifle. The barrel collar appears to be made of a copper-nickel alloy in at least three variations, one of which is a collet-style with longitudinal slits to assist it in flexing outward. Some aluminum barrel collars have been observed. These were most likely manufactured for the Model 1911 than earlier models as aluminum was expensive to manufacture before the end of the first decade of the 20th century. The collar was slightly bigger than the heated and expanded barrel at that point. The barrel collar was captured in milled cuts in the stock nose and front of the hand guard. When the stock and hand guard were clamped down by the barrel bands, the barrel collar prevented either from touching the barrel and interrupting its normal vibration, see Figure 3-48.

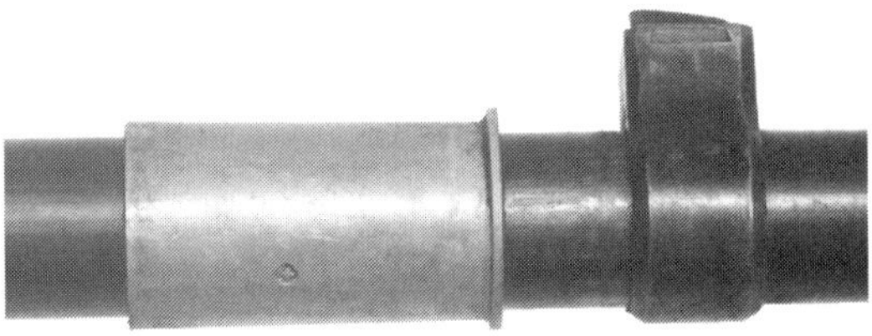

Fig. 3-48. Model 1889 barrel collar. Also used on all rifles and carbines through the Model 1911 series.

Barrel, Model 1911 Rifle

The barrel for the Model 1911 rifle was the same length, 30.7 inches, as the Model 1889 rifle. It was quite similar to the Model 1889 barrel with these exceptions. 1) The lug in front of the rear sight for positioning the hand guard was omitted. 2) The rifling was changed from three grooves to four. The grooves remained 1/6th (0.16 inch) the circumference of the barrel but the lands were narrowed to 1/12th (0.08 inch). Groove depth was reduced from 0.007 inch to 0.005 inch. The nominal diameter of the bore was: top of lands, 0.297 inch, bottom of grooves, 0.307 inch. Most bores actually measure closer to 0.308 inch. 3) The front sight blade was mounted in a dovetail that could be moved left or right in the base by an armorer to zero the rifle for windage, see Figure 3-49. 4) The rear sight base mount was now a ring the same length as the base. It butted against the barrel step and was brazed in place. A dovetail was cut fore and aft and the sight base was driven onto the dovetail and butted against the barrel step, see Figure 3-50.

Fig. 3-49. Model 1911 rifle front sight blade and base. Compare to Figure 3-47 and note that the dovetail projects beyond the base (arrow).

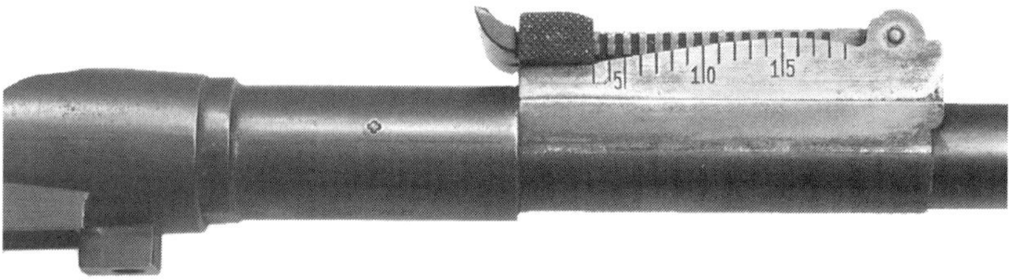

Fig. 3-50. Model 1911 rear sight and base. Note that the base is brazed to the barrel ring.

Barrel, Model 1911 Carbine

The barrel used on the carbine version of the Model 1911 was identical in shape to the rifle barrel but was shorter at 23.3 inches and had a more pronounced taper. The diameter of both the rifle and carbine barrels was 1.115 inches at the breech but the rifle barrel muzzle diameter was 0.591 inch while the carbine barrel muzzle was 0.550 inch in diameter. The Model 1911 *Konstruktion* Bayonet (a.k.a. Model 1914) with the serrated back was designed to fit the Model 1911 carbine and because of the smaller diameter of the carbine muzzle, it cannot be mounted on the rifle. Front and rear sight mountings were the same.

NOTE: Many collectors and others refer to these saw-toothed bayonets as "butcher's bayonets." In fact, they were designed originally to cut through the wooden posts on which barbed wire was mounted in front of trenches. *Konstruktion* troops were "pioneers," which units derived from "sappers." The modern term is "combat engineers." *Konstruktion* troops, or pioneers, were often called upon to clear barbed wire entanglements, but also to cobble together shelters, bridges and bunkers at a moment's notice. These bayonets were excellent saws and while a soldier could defend himself with one, the serrations could have made the bayonet difficult to withdraw from a body.

Barrel, K31 Rifle

The K31 barrel (25.7 inches) was 5 inches shorter than the Model 1911 rifle barrel (30.7 inches) and 2.4 inches longer than the Model 1911 carbine barrel (23.3 inches). The bore diameter, lands and grooves and rifling rate differed slightly from the Model 1911. Nominal bore diameter in the K31 was 0.296 inch (7.51 mm). Groove depth was increased slightly from 0.005 to 0.0055 inch (0.14 mm). Groove diameter remained nominally 0.307 inch (7.79 mm) and the lands were increased from 1/12th (0.08 inch) the bore circumference to 1/10th (0.095 inch).

The primary visual difference between the Model 1911 carbine barrel and the K31 barrel is the single curved step from the breech to the rear of the front sight base in the K31, see Figure 3-51. The rear sight ring was brazed to the barrel and backed against the step. It had a straight taper from the step to the muzzle, 0.860 to 0.562 inch in diameter.

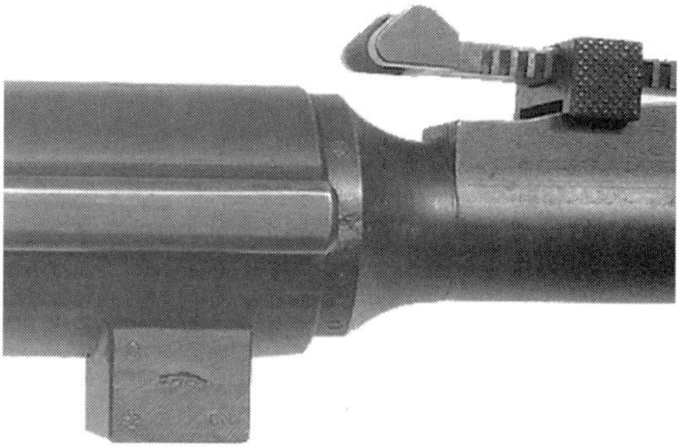

Fig. 3-51. K31 barreled receiver. Note that the K31 barrel has only one step. Note also the recoil lug on the bottom of the receiver.

Front Sight, Model 1889 Rifle

The Model 1889 front sight was mounted on a ring that encircled the barrel and was brazed in place. The top of the ring was a flat platform with a fore and aft dovetail cut across it. The front sight blade had a base with matching cuts that slid into the dovetail, see Figure 3-52. The front sight blade was offset slightly to the left to compensate for bullet drift to the right. The rifle was zeroed for windage at the factory by moving the front sight left or right on its mount on the barrel. Once adjusted, the base of the blade was staked on both sides into the dovetail.

Fig. 3-52. Front sights compared, top: Model 1911; bottom: Model 1889.

Front Sight, Model 1911 Rifle

A new front sight base and blade were designed for the Model 1911 rifle. Like the Model 1889 front sight, the Model 1911 front sight was mounted atop a flat on the front sight ring that encircled the barrel. The Model 1911 front sight differed from the Model 1889 in the fact that the blade was mounted on a base that surmounted the top of the front sight ring. The rifle could be zeroed for windage by moving the sight base in the dovetail. This task was reserved for trained armorers. When the rifle was zeroed, the sight base was staked in place with a punch or chisel. Refer to Figure 3-52 for the difference between the two.

Front Sight, Model 1911 Carbine

Based somewhat on the British Lee Enfield SMLE front sight, it differed considerably in appearance from that used on the Model 1911 rifle, see Figure 3-53. The front sight ring was 0.870 inch long and the front sight blade was dovetailed into it. The dovetail was cut on a diagonal from the front right to rear left to allow the carbine to be zeroed easily for windage at the factory by moving the sight blade forward or backward, which shifted the point of aim right or left. Once adjusted, the base of the blade was staked on both sides into

Fig. 3-53. Model 1911 carbine front sight. Note the diagonal slot for zeroing the sight.

the dovetail. The other difference was the use of "ears" that protected the front sight blade. The ears were 0.77 inch wide at the top.

Front Sight, K31 Rifle

The K31 used the same front sight base and blade as the Model 1911 carbine. The front sight base was enlarged slightly. The ears were 0.78 inch wide at the top, see Figure 3-54. The arrow points to the height marking, a plus (+) sign.

Fig. 3-54. K31 front sight. Compare the protecting ears to those of the Model 1911 carbine in Figure 3-53. Note the curl at the top.

NOTE: Front sights of varying heights were available for the Model 1911 rifle and carbine and the K31 rifle. Front sight heights were: 5.9 mm and marked "-."; 6.2 mm, marked "-"; 6.5 mm (standard, unmarked); 6.8 mm, marked "+"; and 7.1 mm, marked "+.".

Rear Sight, Model 1889 Rifle

The rear sight used on the Schmidt Model 1889 was a redesign of the rear sight used on the Model 1869/71 Vetterli rifle. It was also similar to that used on the Austro-Hungarian Model 1888 rifle, the Dutch Mannlicher and the Italian Carcano. It consisted of a base with high sides, sloped front to back with indentations, or steps, on the left side to correspond to various elevations, see Figure

3-55. The rear sight was graduated from 400 to 2,000 meters in one-hundred-meter increments on the top of the left rail. The left side of the base was marked "300," "500," "1000," "1500" and "2000." The battlesight setting was 300 meters. The sight leaf was hinged at the front on a pin. It was also split with the narrower, left side spring-loaded at the base. By squeezing the left part of the leaf to the right, the leaf could be moved up and down along the slope to set the elevation. The leaf was held at the proper elevation by a lug on the left part of the leaf and grooves on the left, inside of the sight base wall. The eyepiece, or aperture, was mounted on the end of the leaf and had a single "V" notch for sighting.

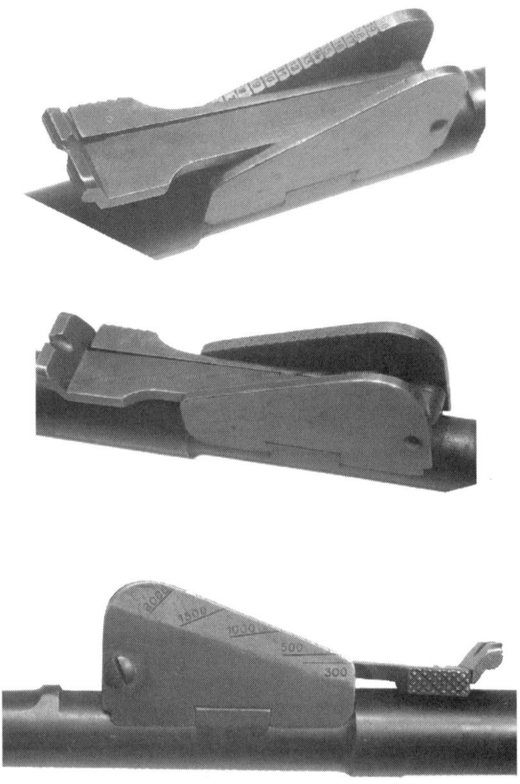

Fig. 3-55. Three views of the Model 1889 rifle rear sight.

The advantage of this sight design lay in the fact that if the rifle were dropped with the leaf raised, the leaf would simply be pushed down into the base, rather than snapped off.

Rear Sight, Model 1911 Rifle

The rear sight used on the Model 1911 rifle and carbine, see Figure 3-56, was similar to the tangent design first developed in Denmark for the Model 1889 Krag rifle and subsequently used on the American Krag rifles as the Model 1898 and Model 1902 rear sights and on the Mauser GEW 98.

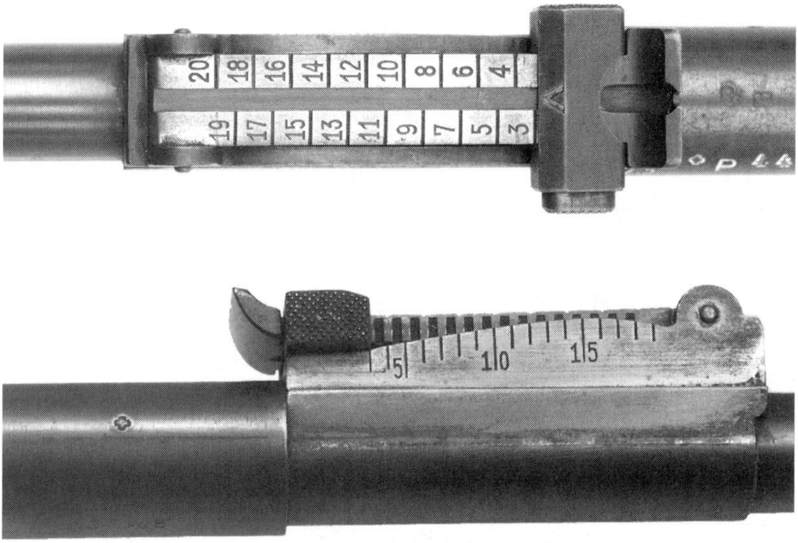

Fig. 3-56. Model 1911 rifle rear sight.

The sight base was mounted on a ring that encircled the barrel. The top of the ring was flat and cut with a fore and aft dovetail into which the rear sight base was inserted.

The rear sight base had a convex curve that coincided with the bullet's trajectory. The leaf was hinged at the front and carried a slide that engaged notches in the bottom right side of the leaf to set the elevation. The slide had a flat top and forward slope. A spring-loaded catch on the left allowed lugs to engage the square detents cut on either side of the leaf. A small inverted "V" was stamped on the forward slope of the slide The aperture bar on the end of the leaf carried a "U" sighting notch.

The base was graduated on the left and right sides, "5," "10," and "15," in one-hundred-meter increments. Two graduation marks preceded the "5" to indicate 300 and 400 meters. The space between the numbers was divided into five one-hundred-meter increments.

The leaf was marked on the top with range graduation marks starting at 300 and extending to 2,000 meters in one-hundred-meter increments. The base was likewise marked on either side with range graduations extending from 300 to 1,900 in one-hundred-meter increments. The range could be set by using either set of range markings on the leaf or base. The rear sight was not adjustable for windage. The parts of the rear sight were stamped with assembly numbers as each part was hand-fitted. A leaf spring in the base pressed against the front of the leaf to hold it in position.

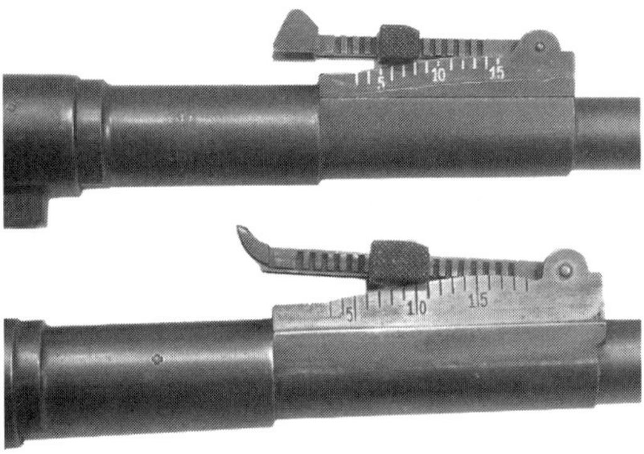

Fig. 3-57. Model 1911 rear sights, top: carbine; bottom: rifle.

Rear Sight, Model 1911 Carbine

The rear sight for the Model 1911 carbine was similar to that used on the rifle, see Figure 3-57. The differences were primarily in the leaf, which had ears protecting the aperture. The leaf was 2.8 inches long and graduated from 300 to 1,500 meters in one-hundred-meter increments. The base was likewise graduated on either side from 300 to 1,500 meters. The protective ears on either side of the aperture bar were triangular in shape, although truncated at the front. They were 0.497 inch high, refer to Figure 3-57, above. The ears protected the sight aperture when the carbine was slung across the back while on horseback or bicycle.

The carbine rear sight was set back farther on the barrel than on the rifle to increase the sight radius.

Rear Sight, K31

The rear sight used on the K31 was quite similar to that used on the Model 1911, see Figure 3-58. The K31 rear sight and mounting ring were made in one piece, i.e., the K31 rear sight base was not mounted in a dovetail on the ring.

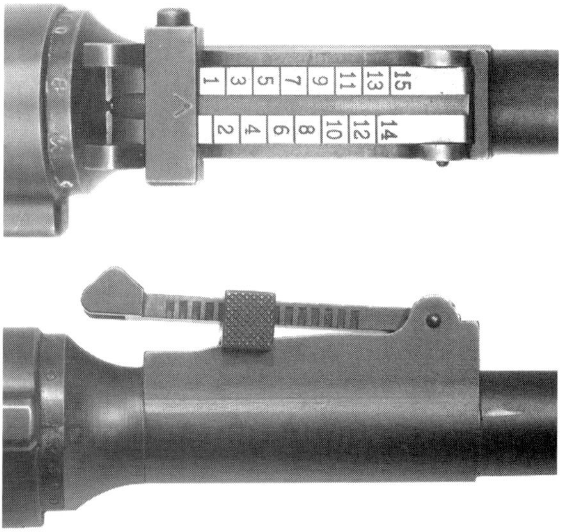

Fig. 3-58. K31 rear sight. Note the similarity to the Model 1911 rear sight, but the lack of range markings on the base.

Range graduations were marked on the Model 1911 sight base but not on the K31 base. The K31 slide was milled flat on top and stamped with a large inverted "V." The protective ears on either side of the aperture were only 0.360 inch high. The graduations on the leaf ran from 100 to 1,500 meters in one-hundred-meter increments. No graduations were marked on the base.

Chapter 4
Sniper Rifles

In some ways it is surprising that the Swiss waited until the advent of World War II to begin developing a sniper rifle. The country's first line of defense is its high and rugged mountains, particularly on its borders with Germany, Austria and Italy–its likeliest adversaries–and sniper rifles are ideally suited for use in mountainous terrain. The shooter can work from an elevated platform and it is harder for the enemy to pursue him. The success in demoralizing enemy troops that snipers on all sides enjoyed in World War I should have served as a warning but as in most other countries, serious work on sniper rifles was ignored in the interwar period. Like most other nations, Switzerland—the home of the League of Nations—was hopeful that the League would prevent future mass military conflicts.

Sniper Rifles

Sniper Rifle ZFKar. 31/42

The ZFKar. 31/42 (*Zielfernrohr-Karabiner)* was the first official sniper rifle adopted by the Swiss military, see Figures 4-1 and 4-2. It was a standard K31 rifle selected for accuracy and equipped with a Kern 1.8x telescopic sight, permanently mounted on the left side of the receiver. The rear sight was the standard tangent sight for the K31 but graduated from 100 to 1,000 meters in one-hundred-meter increments. The telescopic sight was quite unusual in that when used, the front of the scope containing the objective lens was rotated up like a periscope, see Figures 4-3 and 4-4. The reticle pattern was an inverted "V" with right and left pointed sidebars, see Figure 4-5. The advantage of this arrangement lay in the fact that the shooter was able to keep his eye close to the line of sight, which reduced the problem of parallax. Sighting in the rifle and scope was very difficult, however, involving a great deal of trial and error. The ZFKar. 31/42 was approved on July 1, 1944, and some 2,240 were manufactured through 1946 at the Eidgenössische Waffenfabrik, Bern.

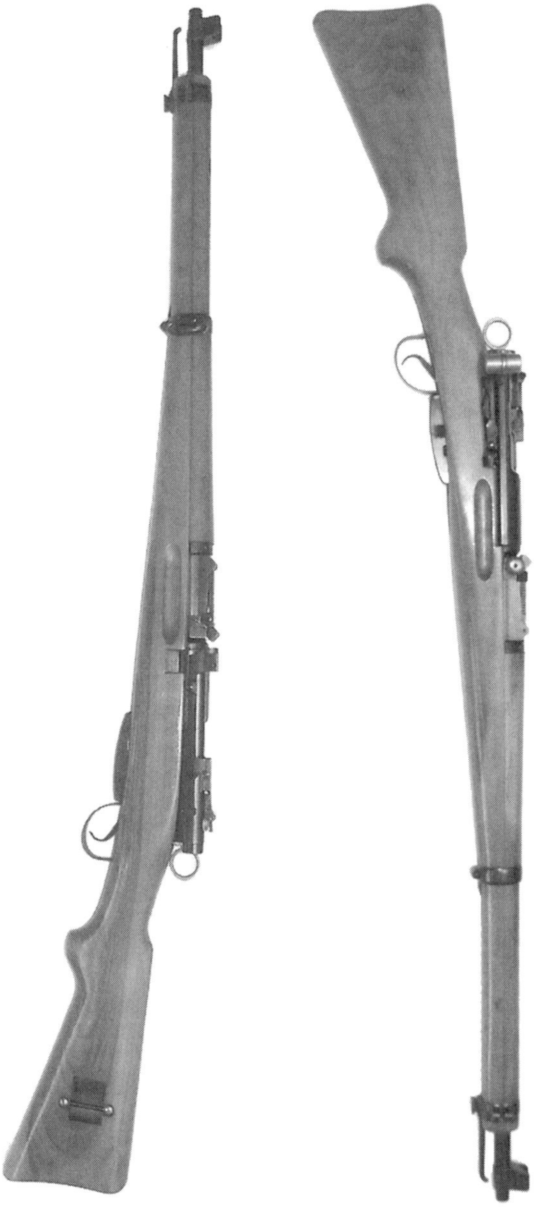

Figs. 4-1 and 4-2. ZFKar.31/42 Sniper Rifle, right and left views. Ian Shein collection.

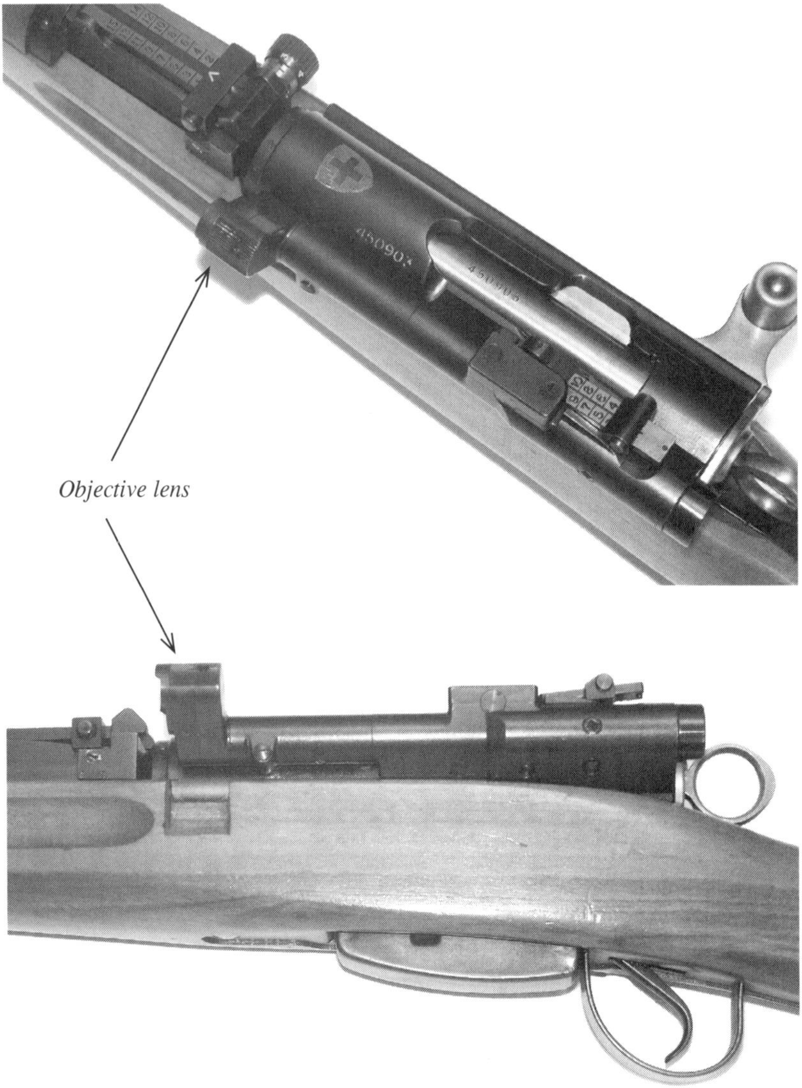

Figs. 4-3 and 4-4 show the front of the Kern 1.8x telescopic sight rotated up for use. The sight tube was mounted close to the receiver with the objective lens raised to reduce parallax. Ian Shein collection.

Sniper Rifle, ZFKar. 31/43

The ZFKar. 31/43 was basically the same sniper rifle/telescopic sight combination as the ZFKar. 31/42 but with a slightly more powerful Kern telescopic sight at 2.8x. Its rear sight was graduated from 100 to 700 meters in one-hundred-meter increments. It is believed that the ZFKar. 31/43 was manufactured as part of, and included in, the same series of 2,240 rifles as the ZFKar. 31/42.

Fig. 4-5. Reticle pattern for the ZFKar. 31/42 and 31/43 telescopic sights.

Sniper Rifle, ZFKar. 31/55

Developed and issued beginning in 1955, the ZFKar. 31/55 was a radical departure from the ZFKar. 31/42 and 43 models, see Figures 4-6 and 4-7, overleaf. The ZFKar. 31/55 was specially designed and built for accuracy. In effect, it was a new rifle having only four parts interchangeable with the K31: the cocking piece, firing pin, firing pin spring or bolt spring and extractor. The stock forend was shortened at the rear barrel band and the pistol grip was checkered. A heavy steel bipod was fastened to the stock, see Figure 4-8, overleaf. An unusual cylindrical, perforated steel flash suppressor and muzzle brake was installed on the muzzle. The unit appears to have been derived partly from the Cutts Compensator.

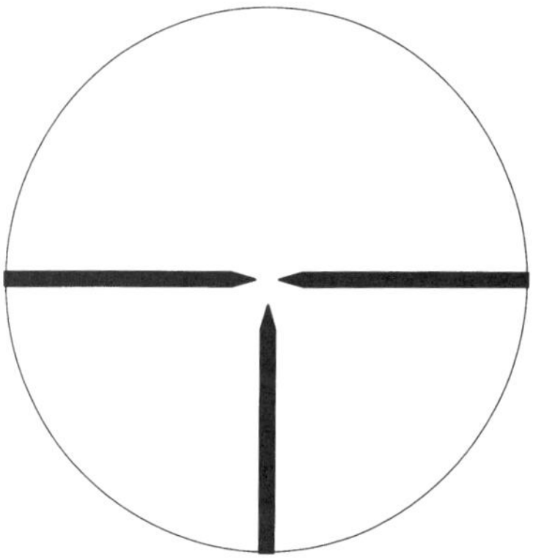

Fig. 4-9. Reticle pattern for the ZFKar. 31/55 telescopic sight.

A new, Kern 3.5x telescopic sight was developed with a range to 800 meters. The reticle was the

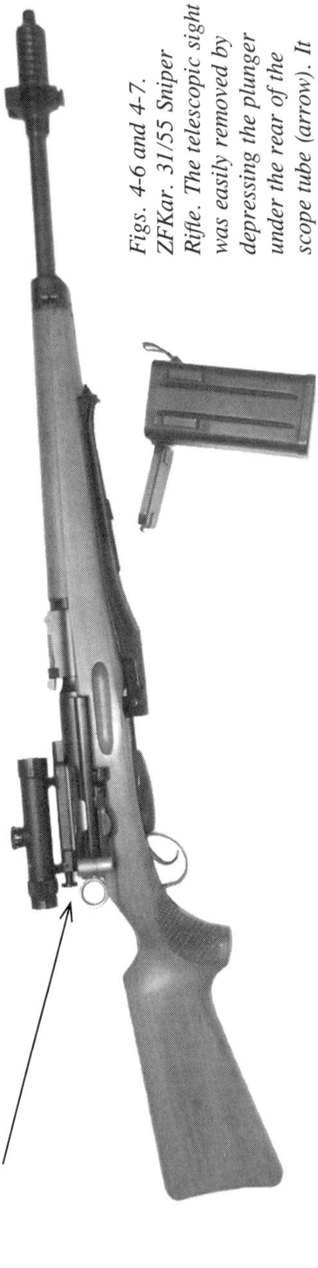

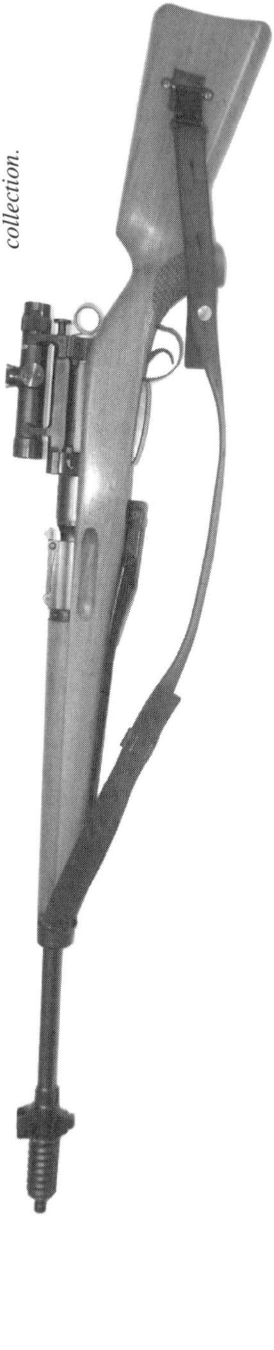

Figs. 4-6 and 4-7. ZFKar. 31/55 Sniper Rifle. The telescopic sight was easily removed by depressing the plunger under the rear of the scope tube (arrow). It was stored in its metal container when not in use. It retained its zero when remounted. Ian Shein collection.

Fig. 4-8. A heavy steel bipod was fastened to the ZFKar. 31/55 Sniper Rifle ahead of the trigger guard. Note the checkered pistol grip. Ian Shein collection.

Fig. 4-10. By depressing the plunger (arrow) on the back of the scope mount, the Kern 3.5x telescopic sight could be lifted off the rifle for storage. The design of the mount allowed the scope to retain its zero setting when remounted.

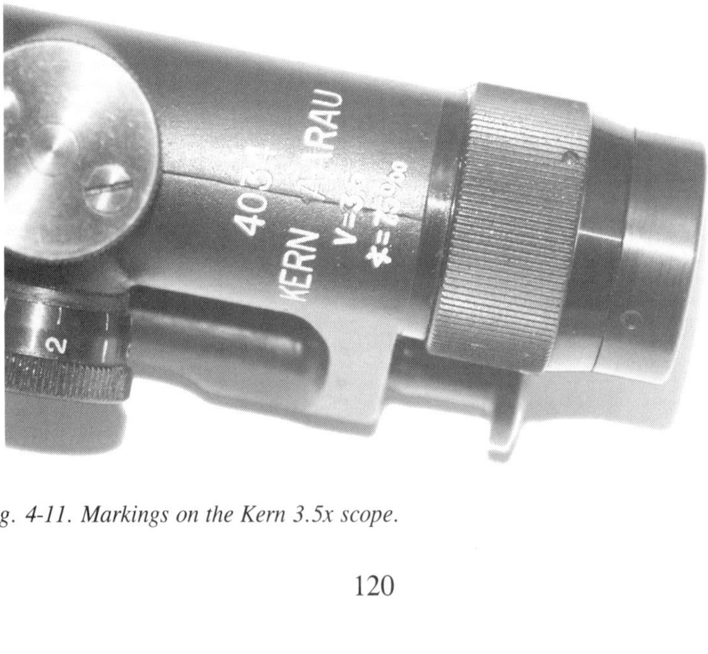

Fig. 4-11. Markings on the Kern 3.5x scope.

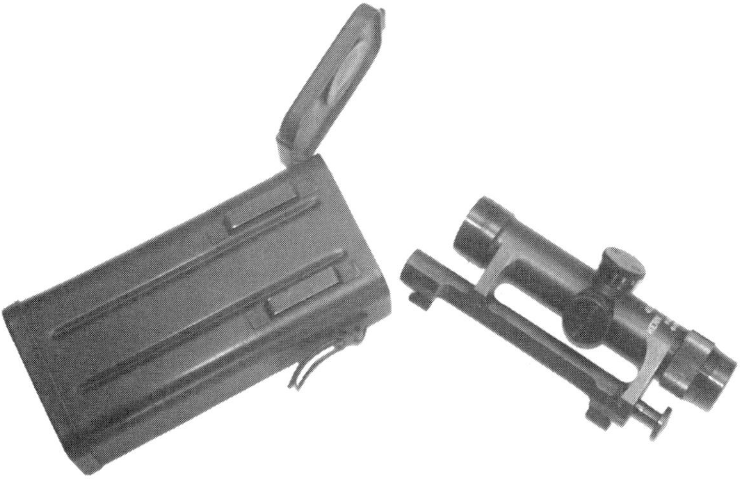

Fig. 4-12. The Kern 3.5x power telescopic sight is stored and carried in its metal container along with cleaning and adjusting tools when not in use. Ian Shein collection.

British pattern using a standing pointed post and pointed right and left side bars, see Figure 4-9. The Kern scope on the ZFKar. 31/55 was attached via two brackets mounted permanently on the left side of the receiver. The scope mated with the brackets using a plunger, see Figure 4-10, arrow. Some 4,150 ZFKar. 31/55 rifles were manufactured between 1957 and 1959 at Eidgenössische Waffenfabrik, Bern.

The scope was marked ahead of the eyepiece, "Serial Number/KERN AARAU/V=3.5/ ∢ = 75 $^{0}/_{00}$,"see Figure 4-11. The telescopic sight and adjusting tool were stored in a metal can when not mounted on the rifle. The can could be worn on the soldier's harness or belt or stored in his pack, see Figure 4-12.

NOTE: In French, the sniper rifle is "*Mousqueton à lunette 31/42 (31/43 or 31/55).*" In German, it is "*Zielfernrohr-Karabiner 31/42 (31/43 or 31/55).*"

Chapter 5
Markings and Serial Numbers

Overview

Markings and types of finish help tell whether or not a Vetterli or Schmidt rifle or carbine is original, has been refurbished by the Swiss military or otherwise changed by previous private owners. Table 5-1 lists all major parts of the Vetterli rifle and carbine, and Table 5-2, the Schmidt rifle and carbine, which are marked with a serial number, inspector's mark or an Ordnance mark.

To understand the difference between original and replacement parts, you must know how parts were made and finished. All parts were manufactured from forged or cast parts which were machined to final shape and dimension in a series of steps. At each step, a gauge made of tool steel was applied to the part by an inspector. The gauge was the exact reverse of the part. If the part did not fit, it was re-machined until it did. When one stage was completed it was passed along to the next stage. When the final shape of the part was achieved and all dimensions were correct according to the gauge, the inspector stamped his initial(s) or mark on the part, or the Swiss cross acceptance mark, or both. The piece was then sent for its final polishing and finish, which would have included tempering and hardening, case-hardening or blueing. The final polishing removed the slight ridge of displaced metal around the periphery of the number, initial or mark.

In addition to the inspector's mark, all major parts will also carry the last three digits of the original receiver number (consult Tables 5-1 and 5-2). If the full or partial serial number is not present on the part, or it does not match, then the part is a replacement.

Significant parts without serial numbers were made as replacement parts. Those parts with full or partial serial numbers that do not match the receiver serial number have been taken from another rifle or carbine.

Table 5-1 Markings on Vetterli Rifle and Carbine Parts			
Part	**Serial Number**	**Inspector's Markings**	**Ordnance Acceptance Mark**
Stock Assembly			
Stock	Y	Y	Y
Forend/Forearm	Partial	N	Y
Band Spring	N	N	N
Butt Plate	N (M1869/71) Y (M1878/81)	Y Y	Y Y
Receiver Assembly			
Receiver	Y	Y	Y
Elevator Lift Assembly	Y	Y	N
Barrel Assembly			
Barrel	Y	Y	Y
Rear Sight Base	Y	Y	N
Rear Sight Leaf	Y	Y	N
Front Sight Blade	N	N	N
Front Barrel Band	Y	Y	N
Middle Barrel Band	Y	Y	N
Rear Barrel Band	Y	Y	Y
Cleaning Rod	N	Y	N
Magazine Assembly			
Tube	N	N	N

Table 5-1, cont. Markings on Vetterli Rifle and Carbine Parts			
Part	**Serial Number**	**Inspector's Markings**	**Ordnance Acceptance Mark**
Follower	N	N	N
Magazine Spring	N	N	N
Bolt Assembly			
Bolt	Y	Y	N
Extractor	N	N	Y
Bolt Sleeve	Y	Y	N
Striker	N	Y	N
Firing Pin	N	Occasional	Y
Main Spring Cover	N (1869/71) Y (1878/81)	N N	Y Y
Bolt Cap	N	Y	Y
Bolt Handle Cam	N (M1869/71) Y (M1878/81)	Y Y	Y N
Trigger Guard Assembly			
Trigger Guard Plate	N (M1869/71)	Y	Y
Trigger Guard	Y (M1878/81)	Y	Y
Trigger	N	Y	Y
Sear Lever	N (M1869/71) Y (M1878/81)	Y Y	N N
Rear Sling Plate	N	N	Y

Table 5-2 Markings on Schmidt Rifle and Carbine Parts			
Part	**Serial Number**	**Inspector's Markings**	**Ordnance Acceptance Mark**
Stock Assembly			
Stock	Y	Y	Y
Hand Guard	Y	Y	Y
Butt Plate	Y (N K31)	Y (N K31)	Y (N K31)
Band Spring	N	N	N
Stock Nose	N	Y	Y
Receiver Assembly			
Receiver	Y	Y	Y
Barrel Assembly			
Barrel	Y	Y	Y
Rear Sight Base	Y	Y	Y
Rear Sight Leaf (both sections)	Y	Y	Y
Front Sight Base	N	N	N
Front Sight Blade	N	N	N
Front Barrel Band	N	N	Y
Rear Barrel Band	N	N	Y
Stacking Rod	N	Y	Y
Collar	N	N	Y
Magazine Assembly			
Magazine	Y	Y	Y
Follower	N	N	Y

Table 5-2, cont. Markings on Schmidt Rifle and Carbine Parts			
Part	**Serial Number**	**Inspector's Markings**	**Ordnance Acceptance Mark**
Spring	N	N	N
Bolt Assembly			
Bolt	Y (M89, 1911) N (K31)	Y N (K31)	Y N (K31)
Extractor	N	Y	Y
Bolt Sleeve	N (M89, 1911) N (K31)	Y	Y
Firing Pin	N	Y	Y
Main Spring	N	N	N
Cocking Piece	N	Y*	Y
Bolt Cap	N	Y	Y
Bolt Handle Cam	N	Y*	Y
Trigger Guard Assembly			
Trigger Guard Plate	Y(1889/1911) N (K31)	N (1889/1911) Y (K31)	Y (1889/1911) Y (K31)
Trigger Guard	N	N	Y
Trigger	Y	Y	Y
Sear Lever	Y (1911 only)	N	Y
Rear Sling Plate			
Magazine Cutoff (M1889 only)	N	N	Y
* Early model 1889 rifles may not show an inspector's mark on this part			

Replacement parts will often be noted with the last three digits of a serial number and a Swiss cross acceptance mark tilted at an angle to the right. Examine the numbers carefully for the furrows of metal around the periphery of the number showing that it was struck after the part received its final polish and finish.

INSPECTION MARKINGS

Swiss Ordnance Department Inspection techniques for the Vetterli and Schmidt rifles were rigorous as could be expected. Every major part of the Vetterli or Schmidt rifle or carbine was tested against carefully machined gauges and stamped with the appropriate stamp indicating that it had passed. The markings are shown in Table 5-3 overleaf, as Figures 5-1 through 5-18.

The Difference Between Cartouches and Stamps

Markings stamped into the metal of Vetterli rifles and carbines are usually in the form of a cartouche. A cartouche in this instance is a circular, oval or rectangular frame enclosing a letter, number or other mark that is stamped into the metal or wood. The cartouches used for the Vetterli created a depression and left the mark raised in the center, see Figure 5-19.

Fig. 5-19. Cartouche, a metal stamp with a letter, number or symbol in relief so that when struck, it left the mark raised in the wood or metal.

In the Schmidt rifles and carbines, most markings were stamped into the metal in the form of simple numbers, letters or symbols with dies machined from tool steel. The exception to the Schmidt markings will be found in the pre-1911 rifles. The chief

Table 5-3
Inspection Markings for Vetterli and Schmidt Rifles

Type of Mark	Mark (Figure)	Found On
Proof Stamp (Beschußstempel)	1 	Schmidt
Shooting Test* (Einschießstempel)	2 	Vetterli
Shooting Test* (Einschießstempel)	3 	Schmidt
Waffenfabrik, Bern Acceptance Mark	4 	K31 (1942–46+)
Proof Mark, Buttstock	5 	Vetterli and Schmidt
Repair Mark	6 	Schmidt
Federal Acceptance Mark*	7 	Vetterli/early Schmidt
Acceptance Mark found on large metal parts	8 	Vetterli
Acceptance Mark found on large metal parts	9 	Vetterli
Acceptance Mark found on small metal parts	10 	Vetterli
Acceptance Mark found on Buttstock	11 	Vetterli
Rejection Mark	12 	Vetterli
Rejection Mark found on metal parts	13 	Vetterli
Rejection Mark found on wooden parts	14 	Vetterli

* Einschießstempel — Swiss ordnance acceptance mark

Table 5-3, cont. Inspection Markings for Vetterli and Schmidt Rifles		
Type of Mark	**Mark (Figure)**	**Found On**
Chrome Nickel Part	15 +CN	K31
Chrome Molybdenum Part	16 +CM	K31
Chrome Vanadium Part	17 +CV	K31
Hämmerli Mark	18 	Schmidt
Sale to Reservist	P, P. Year, P/Year	Schmidt
Repair Marks	○ Last two digits of year Last two digits of year ○	Schmidt
Repair Mark, Waffenfabrik, Bern	WF + last two digits of year	Schmidt
Refurbishment Mark	Last two digits of year	Schmidt
Stock Manufactured By	Roman numerals denoting month and Arabic numerals denoting year	Vetterli and Schmidt found on buttstock

inspector's mark was still stamped as a cartouche.

All markings were stamped before the metal was heat-treated, given its final polish and blued or hardened. If a ridge of raised metal (see Figure 5-20) can be seen around the stamping or cartouche,

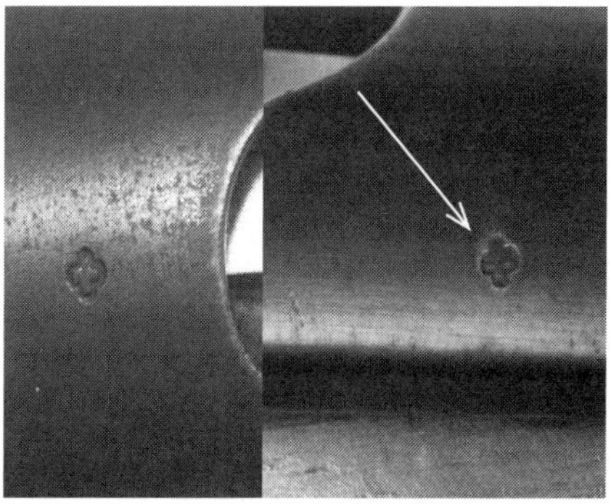

Fig. 5-20. Note the raised ridge of metal around the Swiss cross on the right, denoting that it was stamped after the metal was polished and finished.

then the stamp or cartouche was applied to a finished part which was intended to be used as a replacement. If so, the Swiss cross acceptance mark will usually be tilted clockwise. All inspection markings applied to both Vetterli and Schmidt firearms were stamped using fixtures to hold the die in a specific location. All original inspection markings are always oriented vertically.

NOTE: Stamped marks and cartouches are very small, and their exact size on any one rifle or carbine depends on how deeply it was stamped. The dimensions given here are from rifles and carbines in the North Cape Publications®, Inc., collection and may vary by 0.01 inch or slightly more.

Weapons Marks

Three types of markings are found on all rifles and carbines accepted for military service in Switzerland.

Proof Mark

The first is the proof mark, stamp (*Beschußstempel*), which means that the rifle or carbine met all specifications and was accepted into military service by the Ordnance Department. Three acceptance stamps were in use during the eighty-nine-year period, 1869 to 1958. A cartouche containing a Swiss cross above a flat-bottomed "V," enclosed in a rectangle with clipped corners 0.06 x 0.18 inch, was stamped on the Vetterli and early Schmidt rifles and carbines, refer to Figure 5-7 in Table 5-3. Starting with the Model 1911 Schmidt rifles and carbines, the cartouche was superseded by a stamp of an intertwined "BP" with the "B" reversed, refer to Figure 5-1. From the 1942/46 period to the end of production, it was a "K" and Swiss cross inside a stylized "W" of the factory, "Waffenfabrik" (Weapons Factory) in Bern, refer to Figure 5-4. These marks are found on the receiver and top of the barrel, between the rear sight and receiver ring.

It should be noted that either the "BP" and the Waffenfabrik stamps may be found on those K31s assembled as early as 1942 to

1946, but 1946 is accepted by most collectors as the start date for its use.

Shooting Test Mark

The ordnance acceptance mark is a proof mark indicating that the barreled action had been fired with a proof cartridge, refer to Figures 5-2 and 3 in Table 5-3. This mark is found on both the receiver ring and the barrel between the rear sight and receiver ring. In the Vetterli rifle models, the Swiss cross acceptance mark is a boxed cartouche. In the Schmidt rifles, it is a stamp without a surrounding box.

Principal Federal Inspector Marking

A principal federal inspector was employed by the Ordnance Department to oversee the manufacture and inspection process of small arms. He was always a military officer until 1946 and he (or his designee) stamped all finished and accepted Vetterli and Schmidt rifles from 1869 through 1946 with the initial of his last name. His initial was combined with the ordnance acceptance mark to form the cartouche or stamp.

From 1942 on, all K31 rifle manufacturing was accomplished at the Waffenfabrik in Bern. Sometime between 1942 and 1946, the principal federal inspector's stamp was replaced by facility stamp, the intertwined "BP" with the "B" reversed. But it should be noted that the use of Major Hauri's stamp has been observed on K31 rifles manufactured as late as 1946. He was the last of the military chief inspectors. See Table 5-4 overleaf for a list of principal federal inspectors and their marks, Figures 5-21 through 5-26.

Acceptance Markings

The ordnance acceptance mark (*Einschießstempel*) was also used as an acceptance mark for significant parts. On large metal parts (barrels, bolt bodies, etc.) of the Vetterli, the ordnance acceptance mark was mounted over an inspector's initial in a cartouche, refer to Figures 5-8 and 5-9 in Table 5-3. On small metal Vetterli parts, the ordnance acceptance mark was used alone in cartouche, refer to Figure 5-10. On Schmidt

Table 5-4 Federal Inspector Markings		
Inspector (Kontrolleur)	**Date of Service**	**Example (Figure)**
Major Schmidt	1864 – 1874	21 
Major Werdmüller	1875 – 1879	22 
Major Vogelsang	1879 – 1912	23 
Major Mühlemann	1913 – 1914	24 
Major Hauri	1942 – 1946	25 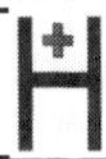
Waffenfabrik, Bern	1942 – 1946 to end of production	26 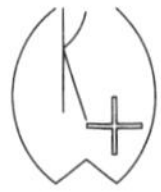

rifle and carbine parts, the ordnance acceptance mark (refer to Figure 5-3) was stamped with a die (no cartouche) on all metal parts as listed in Table 5-2. The cross measured 0.075 by 0.075 inch.

If the ordnance acceptance mark was oriented straight up and down, the part was original. If the cross was tipped to the side, the part was a replacement. This can often be verified by examining the tipped cross carefully to note the ridge of metal around the periphery of the symbol where metal was displaced when the mark was struck (refer to Figure 5-20 for the appearance of this ridge). If the metal part was significant, i.e., barrel, bolt body, etc., you may see the original serial number, or the remains of the original serial number if it was polished out. Any new serial number may be in a different font than that found on other parts or the receiver.

The wooden parts (butt stock, forend and foream) on all rifles and carbines accepted by the Ordnance Department will also show a ordnance acceptance mark and initial inside a shield. If these marks are not present, then the rifle or carbine was manufactured for Cantonal use by police, or paramilitary forces or for civilian sale. Early Vetterli rifles will show the ordnance acceptance mark atop an ornate shield (refer to Figure 5-11 in Table 5-3); later rifles will show the ordnance acceptance mark inside a simple shield outline (refer to Figure 5-5).

NOTE: On early Vetterlis, the ordnance acceptance mark may look more like a "Maltese Cross." This is just an early form.

Parts Inspectors

In addition to the principal federal inspector, as noted above, each individual part was inspected and gauged. If approved, it was stamped with the appropriate acceptance marking as shown in Table 5-3 and depending on the part, the inspector's initial(s). Table 5-5 provides a partial list of various inspectors who were civilian employees of the various contractors and subcontractors.

Table 5-5
Parts Inspector Markings

Initial(s)	Name	Initial(s)	Name
E	Ehrsam	KB	Kaspar Bösch
G	Grunauer	LK	Leonhard Kästli
H	Hauri	LKa	Leonhard Kästli
J	M. J. Genéve	MO	Johann M. Olten
L	Ledermann	MST	Markus Steiger
M	Mechel	MT	Meili Trogen
M	Mühlemann	Ny	Nydegger

Table 5-5, cont. Parts Inspector Markings			
Initial(s)	**Name**	**Initial(s)**	**Name**
O	Oertle	R	Reymann
S	Schenk	RT	Jean Ruckstuhl
V	Vogelsang	Sc	Schläppi
W	Werdmüller	SG	J.J. Schläpfer
AZ	A. Zoller	SL	Seiler Liestal
CB	Camenzind Bouchs	T	Tanner
HO	Hans Oschsenbein	ThK	Thomas Klaus
Hs	Hubschmidt	WS	Walty-Scheumann
JKW	Johann K. Walser	WvS	W. von Steiger

Subcontractor Markings

It was the practice of the Swiss government to spread the manufacture of small arms and parts to private factories throughout Switzerland. The most notable contractor for the Vetterli and the Schmidt Models of 1889 and 1911 was Eidgenössiche Waffenfabrik, in the city of Bern. Their mark was "W+F." This factory later evolved into Waffenfabrik Bern. Another was Schweizerische Industrie-Gesellschaft Neuhausen (SIG). Their mark was a rifle superimposed on the letter "N," or later, just the "N."

A commonly encountered subcontractor is Hämmerli, refer to Figure 5-18 in Table 5-3. Hämmerli was also represented by a hammer in a circle, or by the initials, "Hä."

Other common subcontractor marks are "H&H" (Hämmerli and Hausch, a predecessor of the Hämmerli firm) found on barrels and "U" found on stocks. Table 5-6 provides the names and marks of some, but by no means all, of these factories. The author would appreciate any additional information on this or any other aspect of Swiss small arms by e-mail or letter.

Table 5-6
Parts Inspector Markings

Initial(s)	Name	Initial(s)	Name
AB	Amsler und Blank, Schaffhausen	F&S	Freuler und Sohn, Glarus
A+B	Amsler und Blank, Schaffhausen	GL	Gamma und Inganger, Altdorf
AK	Adolf Knapp, Rheinfelden	Hä	Hämmerli, Lenzburg
ALz	Affolter und Lenk, Zürich	GSW	Gebrüder Sulzer, Winterthur
BS	Burri, Schmid und Compagnie, Luzern	MFA	Munitionsfabrik, Altdorf
B&B	Buholzer und Burri, Luzern	MS	J. Müller und Sohn, Bern
B+B	Buholzer und Burri, Luzern	S.BS	Valentin Sauerbrei, Basel
C	Cordier und Compagnie, Thun	SIG	Schweizerische Industrie-Gesellschaft, Neuhausen
ET	von Erlach und Compagnie, Thun	VS	Valentin Sauerbrei, Basel
Ewy	Escher-Wyss und Compagnie, Zürich	WA	Wahl und Aemmer, Basel
FG	Fabrique d' Armes, Geneva	W&A	Wahl und Aemmer, Basel

Cantonal (State) Markings

Stocks and even major metal parts such as the receiver and butt plate may also show the initials of the Canton, or states, of the Swiss Confederation in which the rifles or carbines were issued to reserve forces, see Table 5-7.

Table 5-7
Cantonal (State) Markings

Initial(s)	Name	Initial(s)	Name
AG	Aargau	KG	Glarus
AI	Appenzell Innerrhoden	KL	Luzern
AR	Appenzell Ausserrhoden	KN	Neuenburg
BE	Bern	NE	Neuchâtel
BL	Basel	NW	Nidwalden
BS	Basel (City)	OW	Obwalden
CF	Freiburg	SH	Schaffhausen
CG	Glarus	SO	Solothurn
CL	Luzern	ST.G	St. Gallen
CN	Neuenburg	SZ	Schwyz
Cth	Thurgau	TH	Thurgau
CV	Vaud	TI	Ticino
CZ	Zürich	UR	Uri
F, also FR	Freiburg	VD	Vaud
GE	Geneva	VS	Valais

Table 5-7, cont. Cantonal (State) Markings			
Initial(s)	**Name**	**Initial(s)**	**Name**
GR	Graubünden	ZG	Zug
JU	Jura	ZZ	Zurich Zeughaus,Zürich
KA	Aargau		

Military Department Markings

Various departments of the Swiss military also marked some of their small arms. The marks known to the author are shown in Table 5-8.

Table 5-8 Military Department and Arsenal Markings	
Initials	**Department**
KMV	War Material Administration
MFA	Munitions Factory, Altdorf
MFT	Munitions Factory, Thun
TA	Technical Department
W+K	Weapons Control (Inspection)

Miscellaneous Markings

From the Model 1911 to the end of production, many stocks will show the date of their manufacture on the end of the buttstock under the butt plate. The month will be marked in Roman numerals, and the year in Arabic numerals. Keep in mind that as much as a year or two could pass between the time the stock was manufactured and the time it was mated to a rifle or carbine.

K31 bolts will show the date of manufacture. This is usually stamped on the bolt cap after the initials of the subcontractor. It is not an infallible way of dating a rifle as both the bolt cap and bolt could have been changed in the past. But if the serial number on the bolt sleeve matches the receiver serial number and they fall into the same year as the bolt cap date, then it is a fairly safe bet that the bolt is original to the rifle.

Finally, various military units applied their own markings in the form of regimental and company numbers and letters. Rack numbers are often observed as well. These last were applied to weapons which remained in a central armory and identified their location within the armory weapons rack.

Serial Numbers

Serial numbers in the Vetterli series appear to run consecutively from start to end of production. The author knows of no reliable list of serial numbers and because of the relative scarcity of Vetterli rifles in the United States, has not been able to compile a serial number list with any degree of confidence.

Vetterli series numbers appear to have been assigned consecutively from 1 through circa 230,000 or more. Those *S tutzer* rifles and carbines newly manufactured and not rebuilt from existing rifles, i.e., the Model 1871 Type 1 and 2 carbines, were numbered in their own serial ranges. All others retained the receiver serial number of the rifle from which they were rebuilt.

Determining the manufacture date of Schmidt rifles produced from 1889 through 1958 depends entirely on serial number ranges with only two exceptions—Model 1889/90 short rifles and Model 1905 carbines rebuilt as Model 1911 carbines.

The Model 1889 rifle was made in four variations: the basic Model 1889, the Model 1889 with changes of 1896, the Model 1889 with changes of 1900, and the Model 1905 carbine. But each model variation was a separate production run and no Model 1889 rifles were converted to the Model 1889/96 nor were any Model 1889s and 1889/96 rifles converted to Model 1889/00. The same holds true for the Model 1911 rifle and the K31.

As noted above, some Model 1911 carbines were rebuilt from

Model 1889/00 short rifles and Model 1905 carbines. Serial numbers for the former range from circa 1 to 18,750. For the latter, they range from circa 20,001 to 27,900.

Those Model 1911 carbines built from Model 1889/00 short rifles will show the original serial number of the earlier rifle on its receiver.

Those Model 1911 carbines rebuilt from Model 1905 carbines will have the number "2" stamped ahead of the serial number plus sufficient zeroes as needed, i.e., a Model 1905 carbine with an original serial number of "5678" will have had added a "2"; a Model 1905 carbine with an original serial number of "345" will have had added a "2" and "0." An original Model 1905 carbine with a serial number of 23 will have had added a "2," "0" and "0," etc.

It is imperative that the collector understand that a Model 1911 carbine with a serial number within either of those two ranges actually started life as a Model 1889/00 short rifle or a Model 1905 carbine.

A further hint. The serial number of a Model 1911 carbine built from a Model 1905 carbine will have a serial number composed of two different fonts (styles). Examine the leading numbers closely. If the first one, two, or three digits are in a different font from the rest of the number and slightly misaligned, then your Model 1911 carbine began life as one of 7,900 Model 1905 carbines!

Finally, all true Model 1911 carbines, i.e., Model 1911 carbines manufactured originally as Model 1911 carbines, will have serial numbers in the range 30,001 through 215,100.

In addition to the arms manufactured for the Swiss military, rifles and carbines were manufactured for various other Government departments and police forces and for private sale. These were all numbered in their own serial number ranges. Schmidt rifle and carbine serial numbers are shown in Table 5-9. Table 5-10 presents a list of serial numbers for these weapons which were mostly manufactured for private sale and use. The serial number tables are from *Die Repetiergewehre der Schweiz* by Reinhart, Sallaz and am Rhyn with slight amendments.

Table 5-9 Schmidt Serial Numbers	
Model 1889 Repeating Rifle	
1891	1-24,900
1892	24,901-115,000
1893	115,001-174,100
1894	174,101-174,887 175,001-183,000
1895	183,001-195,500
1896	195,501-208,000
1897	208,001-212,000
Model 1889/1896 Repeating Rifles	
1895	1-50
1897	212,001-221,000
1898	221,001-236,500
1899	236,501-252,000
1900	252,001-268,500
1901	268,501-280,000
1902	280,001-292,000
1903	292,001-302,000
1904	302,001-310,000
1905	310,001-316,000

Table 5-9, con't. Schmidt Serial Numbers	
Model 1889/1896, continued	
1906	316,001-322,000
1907	322,001-327,000
1908	327,001-332,000
1909	332,001-336,800
1910	336,801-343,100
1911	343,101-345,000
1912	345,001-349,000
Model 1889/1900 Short Rifles	
1901	1-5,000
1902	5,001-5,885
1903	5,886-6,800
1904	6,801-7,800
1905	7,801-9,300
1906	9,301-10,800
1907	10,801-12,550
1908	12,551-14,300
1909	14,301-15,950
1910	15,951-17,450
1911	17,451-18,750

Table 5-9, con't. Schmidt Serial Numbers	
Model 1905 Cavalry Carbine (Kavalleriekarabiner)	
1906	1-810
1907	811-3,600
1908	3,601-6,200
1909	6,201-6,700
1910	6,701-7,300
1911	7,301-7,900
Model 1896/1911 Infantry Rifles (Infanteriegewehre)	
These rifles were converted from Model 1889/1896 rifles between 1912 and 1920 and are all numbered in that range.	
Model 1911 Infantry Rifle (Infanteriegewehre)	
1912	355,001-361,000
1913	349,001-355,000 361,001-369,000
1914	369,001-374,000
1915	374,001-396,000
1916	396,001-428,000
1917	428,001-459,000
1918	459,001-479,000

Table 5-9, con't. Schmidt Serial Numbers	
1919	479,001-482,000
Model 1911 Carbine (Karabiner)	
1914	30,001-36,000
1915	36,001-40,000
1916	40,001-53,000
1917	53,001-54,400 55,001-62,000 62,301-70,000
1918	70,001-89,000
1919	89,001-91,000
1920	91,001-97,500
1921	97,501-10,600
1922	10,601-113,500
1923	113,501-121,400
1924	121,401-129,000
1925	129,001-136,500
1926	136,501-146,400
1927	146,401-159,000
1928	159,001-170,500
1929	170,501-183,300
1930	183,301-193,200

Table 5-9, cont. Schmidt Serial Numbers	
1931	193,201-202,000
1932	202,001-211,200
1933	211,201-215,100
Model 1911 Carbines Converted from Model 1889/1900 Short Rifles and Model 1905 Carbines	
The dates of these conversions are not reliably known but are thought to have occurred between 1913 and 1920.	
K31 Carbine (Karabiner)	
1933	520,010-521,202
1934	521,203-536,736
1935	536,737-550,400
1936	550,401-561,727
1937	561,728-573,366
1938	573,367-583,700
1939	583,701-599,000
1940	599,001-632,575
1941	632,576-686,725
1942	686,726-736,075
1943	736,076-786,550
1944	786,551-838,450

Table 5-9, cont. K31 Serial Numbers	
1945	838,451-864,650
1946	864,651-880,250
1947	880,251-901,200
1948	901,201-921,300
1949	921,301-936,800
1950	936,801-950,000
1951	950,001-973,050
1952	973,051-994,450
1953	994,451-999,999 215,001-222,450
1954	222,451-239,600
1955	239,601-250,850
1956	250,851-257,250
1957	257,251-260,200
1958	260,201-263,330

Table 5-9, con't. Schmidt Serial Numbers
Model 1931/1942 & 1943 Sniper Rifle (Zielfernrohr-Karabiner)
Unknown
Model 1955 Sniper Rifle (Zielfernrohr-Karabiner)

Table 5-9, con't. Model 1955 Serial Numbers	
1957	1,001-1,170
1958	1,171-4,200
1959	4,201-5,000
1960+	5,001-5,150
Serial number tables adopted from *Die Repetiergewehre der Schweiz,*" Verlag Stocker-Schmid, Dietikon-Zurich, 1991	

Table 5-10 Schmidt Non-Military Serial Numbers	
Model 1889 Repeating Rifle	
1891	E 1-39
1895	E 43 and 45
1896	E 42 & 44
Used for training	
Model 1889/1896 Repeating Rifles	
1897	E 2 & 3
1897	E 4-7
1900	E 8-11
1904	E 12-15
1908	E 16-19

Table 5-10 Schmidt Non-Military Serial Numbers	
Used for training	
Model 1889/1900 Repeating Rifle	
1901/02	P 1-176
1903	P 72-150
1904	P 174-472
	P 501-550
	P 551-944
	P 951-1140
	P 1141-1243
Private sale	
Model 1911 Repeating Rifle	
Years not known	P 500-5200
	P 5219
	P 5251-5948
Private sale	
Model 1911 Carbine	
1913-1919	P 2001-2300
	P 2013-2030
	P 2037-2044
	P 2071-2100

Table 5-10, cont.	
Model 1911 Carbine, cont.	
1913-19, cont.	P 2101-2200
	P 2201-2204
	P 2205-2223
	P2251-2300
Private sale	
Model K31, Training	
1947	E 519767-519796
Model K31, Private Sale	
1934	Unknown
1935	400001-400500
1936	401001-401100
1937	401101-401250 400501-400700
1938	400701-400850
1939	401501-401520
1940	400851-401000 401521-402050 402151-402200 402251-402300
1941	402301-402500

Table 5-10, cont.	
Model K31, Private Sale, cont.	
1942	402501-403150
1943	403151-403500
1944	403501-403900
1945	403901-404500
1946	404501-405337
1947	405322-406250 E 519767-519796
1948	406251-406650
1949	406651-407300
1950	407301-407700 408001-408050
1951	407701-408000 408051-408100
1952	408101-408400
1953	408401-408850 409051-409100
1954	408851-409050 409101-409200
1955	409201-409650 410151-410250
1956	409651-410100

Table 5-10, cont.	
K31, Private Sale, cont.	
1957	410101-410150 410251-410550
1958	410551-411025
1959	411026-411450
1960	411451-411750
1961	411751-412050
1962	412051-412450
1963	267331-267380 412451-412850
1968-69	Unknown
1971	269431-269580 269881-269930
1972	Unknown
Jubilee Model K31	
1981	001-500

Serial Numbers in Practice

As a general statement, serial numbers on the receiver and various other metal parts should match in font style and size on original Vetterli or Schmidt rifles and carbines with only those M1911 carbines manufactured from existing Model 1905 carbines as the exception. In practice this may not be the case.

During refurbishment, replacement parts may have been renumbered with the receiver's serial number. For instance, the author owns both a Model 1889 and a Model 1911 rifle with several parts that were clearly stamped with the last three digits of the respective receiver's serial number after the part was polished and blued. On these parts, the ordnance acceptance mark was also stamped, but after the part was polished and finished. The cross was tipped to the right rather than standing vertical.

Markings by Part

Receiver, Vetterli

The Vetterli receiver will show the manufacturer's name and address (city only) on the left side. The serial number was stamped above the manufacturer's names on the Model 1869/1871 receiver and between the city and model year on the Model 1878/1881 receiver. The Model 1869/1871 receivers do not show a model year (refer to Figure 1-37). Model 1878 and 1881 receivers will be marked either "M78" or "M81." Refer to Figures 1-22 and 1-45.

The Swiss ordnance acceptance mark will be found on the receiver ring and barrel.

Various inspection markings will be found on the left side of the upper tang and on the side flat of the lower tang between the two rear screw holes.

Fig. 5-27a. Model 1889 Schmidt barreled receiver showing the location of the serial numbers, proof and acceptance stamps.

Receiver, Schmidt

The receiver will show a number of markings, see Figure 5-27a.

1) The serial number was stamped on the left side just above the line of the stock in all models.

2) An ordnance acceptance mark was stamped on the top of the receiver ring. If the rifle has undergone extensive refurbishment or been modified to a new model (i.e., conversion from a Model 1889/00 to Model 1911 carbine), it will show a second ordnance acceptance mark on the right side of the receiver ring. The second ordnance acceptance mark will usually be tipped to the right.

3) A variety of inspector markings and ordnance stamps will be observed on the underside of the receiver.

Fig. 5-27b. Vetterli and early Schmidt acceptance cartouche on the barrel and ordnance acceptance mark on the receiver.

4a) When the barreled action of the Model 1889 rifle or carbine was proofed, it was stamped on top of the receiver with a cartouche consisting of a "V" surmounted by the ordnance acceptance mark, see Figure 5-27b; refer to Figure 5-2 in Table 5-3 and also refer to Tables 5-1 and 5-2 for the differences between the Vetterli and Schmidt proofs. The intertwined "BP" proof will usually be found on the barrel.

4b) Model 1911 and K31 receivers manufactured to 1942–1946 show the intertwined "BP" proof mark on the receiver ring, refer to Figure 5-1, Table 5-3. From 1944-1946 to the end of production the K31 receiver will show the "K" and ordnance acceptance mark in the stylized "W" proof of the Waffenfabrik, Bern, refer to Figure 5-4, Table 5-3. The "BP" mark has also been observed on these receivers.

5) If a military rifle or carbine was sold to a Swiss citizen leaving the reserves, it will be hand stamped with a large "P" for "*Privat*," meaning private sale. The mark was usually found under or beside the serial number on the left side, see Figure 5-28. The "P" mark may

Fig. 5-28. A "P" marked adjacent to the serial number indicates that the rifle or carbine was sold to the Swiss soldier when he left active duty or the reserves. The caliber, "7.5" millimeter, was added to the top example.

sometimes be accompanied by a two-digit number designating the year in which the sale was made.

6) The right side of the K31 forward receiver lug will show the ordnance acceptance mark, an inspector's mark, the subcontractor's symbol or initials and two initials indicating the alloy from which the receiver was forged.

K31 receivers and other parts will also show the initials, "CN" for Chrome Nickel, "CM" for Chrome Molybdenum and "CV" for Chrome Vanadium alloy of steel.

7) The underside of the receiver and tang presses against the bearing point of the trigger bears, was polished smooth and left bright. If this area was blued, the receiver has been refinished, Figure 5-29.

Stock and Hand Guard, Vetterli

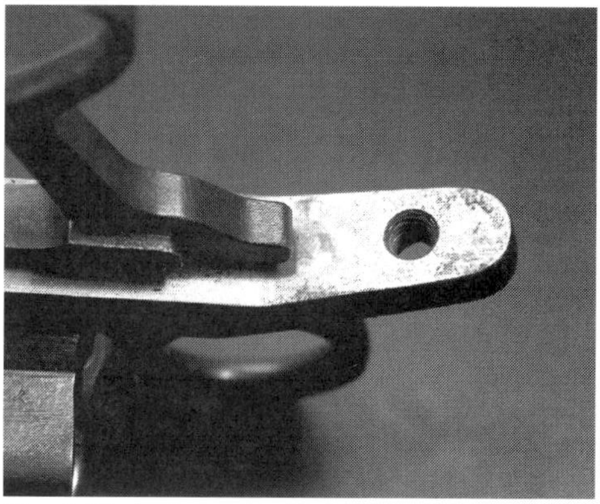

Fig. 5-29. If the underside of the tang is bright, the receiver probably has not been refinished.

1) The receiver serial number was stamped vertically on the lip of the buttstock that enters the receiver.

2) The ordnance acceptance mark/shield was stamped on the right side of the buttstock.

3) An ordnance acceptance mark was stamped on the top of the comb just ahead of the butt plate.

4) The ordnance acceptance mark/shield with an inspector's initial was stamped in the tang cutout in the Model 1878/1881 only.

5) The principal federal inspector's initials in script was stamped on the top of the comb just ahead of the butt plate tang.

6) A lozenge containing inspection numbers or letters was stamped between the principal federal inspector's initials and the butt plate tang. In the case of subcontracted parts, this will be replaced by a similar marking containing the initials of the subcontracting firm.

7) Military unit markings *may* also be found on the buttstock.

8) The Model 1869/1871 forend carries the last three digits of the receiver serial number and the cross/shield ordnance mark on the right side behind the rear band position. No markings have been observed in the barrel channel. The nose of the forend/forearm will show various inspector's marks and symbols.

9) The Model 1878/1881 forend/forearm carries the last three digits of the receiver serial number on the left side of the rear lip that fits into the front of the receiver. The cross/shield with an inspector's initial was stamped on the right side behind the rear band position. No markings were observed in the barrel channel.

Stock and Hand Guard, Schmidt

1) The barrel channel carries the receiver serial number in all models, see Figure 5-30.

2) Various inspector markings (single or double initials or numbers) will be found stamped in the barrel channel, the receiver inletting and in the inletting for the rear sling swivel, see Figure 5-31.

3) Variations include a circle with the inspector's initials and the ordnance acceptance mark, see Figure 5-32.

4) The stock may also carry various symbols such as a cartouche of a rifle superimposed on the letter "N," or the "N" alone indicating manufacture by SIG, a subcontractor, see Figure 5-33. Another common stock symbol was a hammer in a circle for the firm of Hämmerli.

5) The right side of the buttstock was marked with the ordnance acceptance mark inside a shield inspection marking, see Figure 5-34.

6) The hand guard will also carry the receiver serial number, stamped on the underside. K31 hand guards are marked on the underside with the ordnance acceptance mark and a contractor's mark, see Figure 5-35.

Butt Plate, Vetterli

Vetterli butt plates will show the ordnance acceptance mark on the tang. On the inside, will often be found molded into the plate, the subcontractor's initials or code.

The Model 1878/1881 butt plate will also show the last three digits of the receiver serial number.

Butt Plate, Schmidt

The serial number, inspector's markings and the ordnance mark will be found stamped on the butt plate tang of the Model 1889 rifles, see Figure 5-36.

On the Model 1911 butt plate, the Swiss inspection cross was stamped on the rudimentary tang and the last three digits of the serial number are stamped on the inside of the plate, near the toe.

K31 butt plates are unmarked.

Fig. 5-30. Schmidt rifle barrel channel with serial number.

Fig. 5-31. Inspector's markings stamped in the Schmidt barrel channel.

Fig. 5-32. Stockmaker's marking and the ordnance acceptance mark.

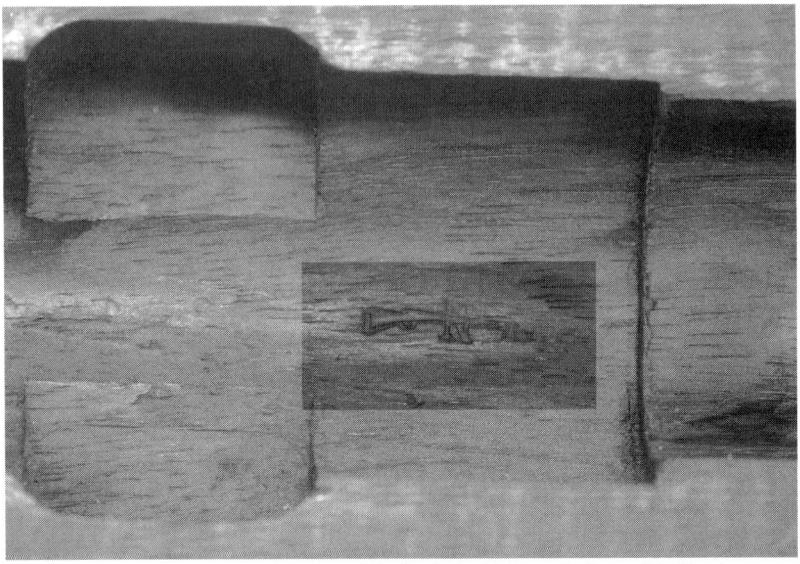

Fig. 5-33. Contractor's mark stamped in the Schmidt barrel channel indicating manufacture by SIG.

Fig. 5-34. Ordnance acceptance mark/shield stamped on the right side of the Schmidt rifle or carbine buttstock.

Fig. 5-35. Schmidt hand guard markings, above: serial number; below: inspector's marking and ordnance acceptance mark.

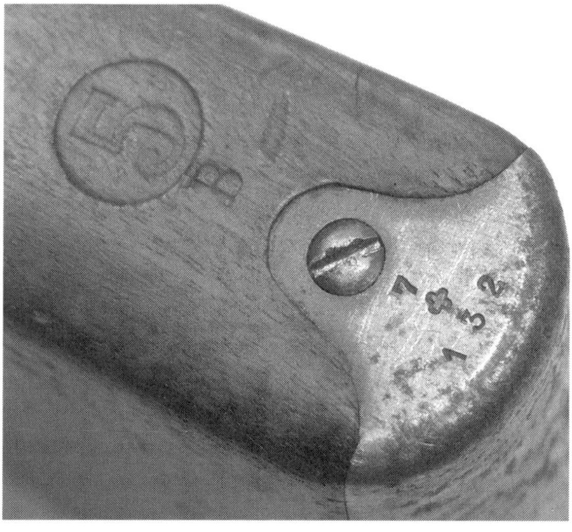

Fig. 5-36. Schmidt buttstock and butt plate markings: buttstock markings include a military unit or rack number and inspector's mark. The butt plate carries an inspection mark, the ordnance acceptance mark and the last three digits of the receiver's serial number.

Trigger Guard and Magazine Assembly, Vetterli

1) The trigger guard plate will show inspector's markings and the ordnance acceptance mark.

2) The trigger will show an inspection marking.

3) The Model 1869/71 sear lever will show an inspection marking only. The Model 1878/81 sear lever will show the last three digits of the serial number and an inspection marking.

Trigger Guard and Magazine Assembly, Schmidt

1) The trigger guard will show the last three digits of the receiver serial number on the inside of the trigger guard plate in all models but the K31, see Figure 5-37.

2) The ordnance acceptance mark was stamped on the outside of the Model 1889 and Model 1911 trigger guard plates, see Figure 5-38.

3) The K31 trigger guard plate has the ordnance acceptance mark and inspection mark stamped on the inside, just ahead of the magazine well, refer to Figure 5-37.

4) Trigger guard bows on the Model 1889 and Model 1911 are show the ordnance acceptance mark, see Figure 5-39.

5) The Model 1889 trigger was marked on the right side of the finger lever with the last three digits of the serial number and an inspection mark, see Figure 5-40.

6) The Model 1911 trigger was marked on the right side of the finger lever with the last three digits of the serial number. The sear lever was also so marked, see Figure 5-41.

Magazines, Vetterli

The magazine tube, follower and magazine spring are all unmarked.

Magazines, Schmidt

1) Schmidt magazines for the Model 1889 are marked with the ordnance acceptance mark, inspector's marks and a subcontractor's mark, see Figure 5-42.

Fig. 5-37. Trigger guard plates with markings, l-r: K31, Model 1911 and Model 1889.

Fig. 5-38. Exterior, trigger guard plate.

Fig. 5-39. Model 1889 and 1911 trigger guard bow.

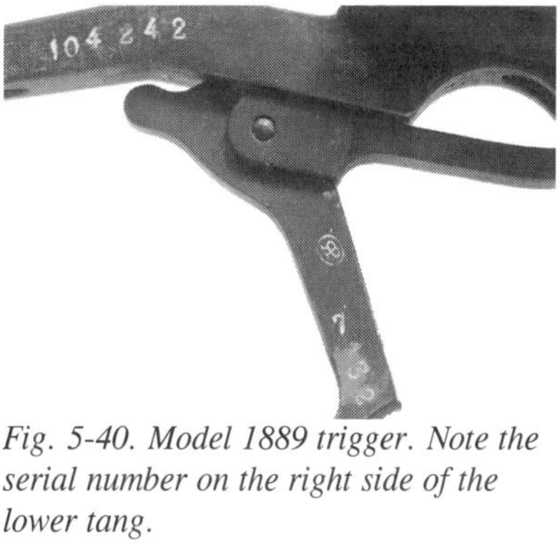

Fig. 5-40. Model 1889 trigger. Note the serial number on the right side of the lower tang.

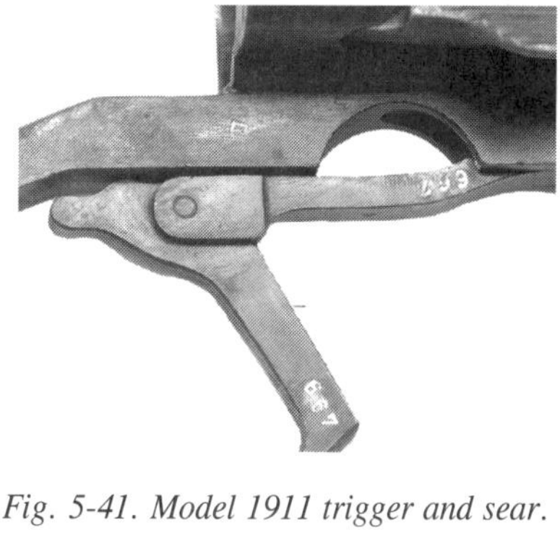

Fig. 5-41. Model 1911 trigger and sear.

2) The full receiver serial number was stamped on the outside of all Schmidt magazines. The serial number was stamped on the left side of the Model 1889 magazine, see Figure 5-43, and on the bottom of the Model 1911 and K31 magazines, see Figure 5-44.

Magazine Release

Only the Model 1889 magazine cutoff switch was marked with the ordnance acceptance mark, see Figure 5-45. The magazine releases on the Model 1911 and K31 magazine are unmarked.

Bolt Release, Schmidt Only

The bolt release for the Model 1889 rifle was marked on the bottom with the ordnance acceptance mark and an inspector's initial. The Model 1911 bolt release was only marked on the bottom with the ordnance acceptance mark and the K31 was unmarked, see Figure 5-46.

Barrel, Vetterli

The Vetterli barrel will show the serial number on the second flat on the left side. The other flats will show a variety of inspector's initials, numbers and symbols. The round bottom of the barrel beneath the chamber will also show a variety of stamped inspector's initials. Cartouches consisting of boxed initials of inspectors will be found on any of the flats and on the rounded bottom. If the barrel was manufactured by a subcontractor, those initials will be stamped in this area, usually in letters two to three times the size of the factory inspectors. Subcontractor-manufactured barrels will often show a three- or four-digit number, which may indicate a manufacturing lot.

Barrel, Schmidt

The barrel will show the receiver serial number on the left side, just above the stock line, see Figure 5-47. It will be visible in the Model 1889 and 1911 rifle but not the Model 1911 carbine as the hand guard extends over the receiver ring. The K31 barrel was stamped with the

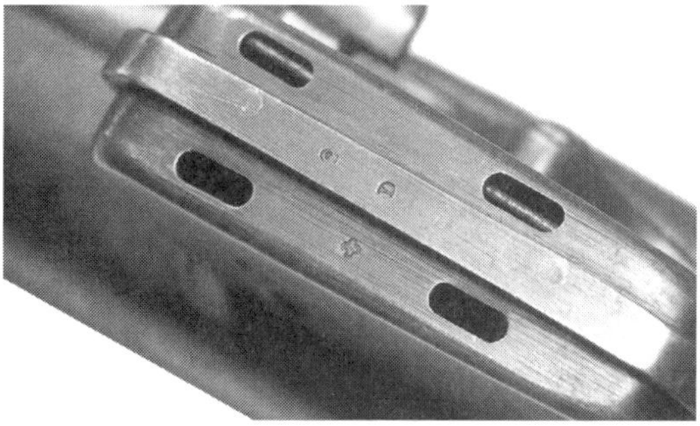

Fig. 5-42. Model 1889 magazine showing inspection markings and ordnance acceptance mark.

Fig. 5-43. Model 1889 magazine showing serial number on the left side.

Fig. 5-44. Top to bottom: Model 1889, M1911 and K31 magazines showing markings.

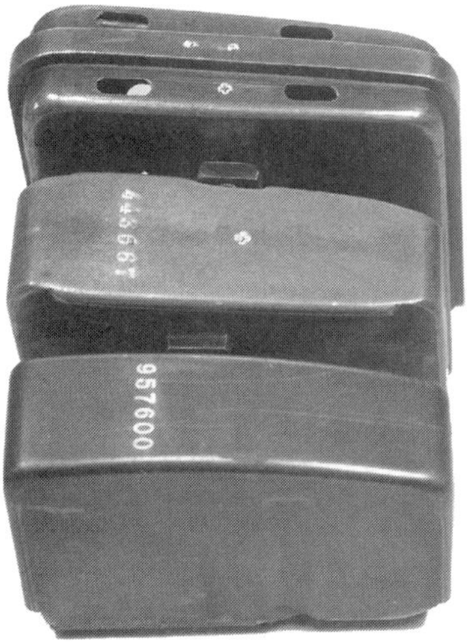

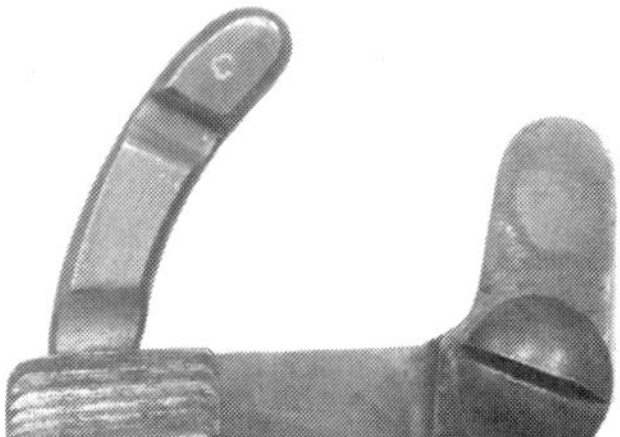

Fig. 5-45. Model 1889 magazine cutoff switch is marked only with the ordnance acceptance mark.

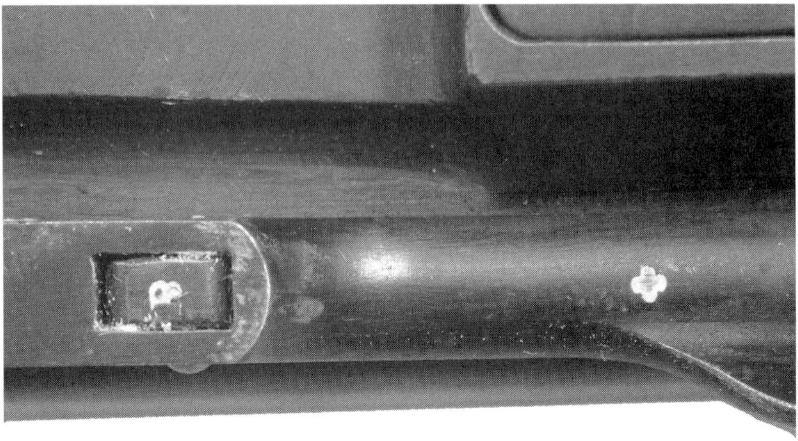

Fig. 5-46. The Models of 1889 and 1911 bolt releases are marked on the bottom with the ordnance acceptance mark only; the K31 bolt release is unmarked. The inspector's initial, visible on the left, is stamped onto the receiver stud on both the Models of 1889 and 1911 but not the K31.

Fig. 5-47. Schmidt barrels (all models) will show the receiver serial number (1), the firing proof (2), index mark (3), ordnance acceptance mark on the receiver (4). The serial number is not visible on the Model 1911 carbine as it is covered by the extended carbine hand guard. See Figs. 5-49 – 5-52 for K31 markings.

last three digits of the receiver serial number on the first step at the 9 o'clock position.

The acceptance mark will be stamped on the barrel behind the rear sight in all except the K31, where it will be found on the periphery of the barrel just ahead of the receiver ring.

A variety of inspector markings will be found on the underside of the barrel near the breech end as well as contractor markings and one or more fitting numbers, see Figure 5-48.

Each time the barrel was proofed, it was stamped with the ordnance acceptance mark. More than one "ordnance acceptance mark" indicates that the barrel was used either as a replacement or it was used in a conversion to the new model, i.e., Model 1889/1900 to Model 1911 carbine or Model 1905 carbine to Model 1911 carbine. Additional crosses should be tipped to one side.

Fig. 5-48. Model 1911 barrel showing fitting and contractor's marks on the underside of the barrel. H&H indicates manufacture by Hämmerli & Hausch.

The intertwined "BP" proof will usually be observed on the barrel.

The barrel collar will show the ordnance acceptance mark in all weapons in which it was used. To summarize:

Model 1889 barrels will show the:

1) serial number at the 10 o'clock position.

2) firing proof mark and barrel index mark at 12 o'clock.

3) ordnance acceptance mark and ordnance proof mark at 3 o'clock.

4) subcontractor and subcontractor's inspection marks, fitting numbers, factory inspection marks and other marks at 6 o'clock.

Model 1911 barrels will show:

1) ordnance acceptance mark and serial number, all at the 9 o'clock position. Also, the caliber designation and "P" for private sale mark (if any) will be stamped before or below the serial number.

2) Firing proof (Reversed B joined with P), inspector's mark and Ordnance inspector's mark at the 12 o'clock position.

3) ordnance acceptance mark at 3 o'clock.

4) Fitting number and subcontractor's inspection and identification marks at the 6 o'clock position.

K31 barrels markings are stamped around the periphery of the first barrel step and include, starting at the 12 o'clock position, the intertwined BP firing proof mark, last three digits of the serial number, ordnance acceptance mark, fitting number, subcontractor's mark and other inspection marks.

Just ahead of the rear sight band will often be found an inspection number or letter stamped into the barrel at the 6 o'clock position, see Figures 5-49, 5-50, 5-51 and 5-52.

Rear Sight Assembly Markings, Vetterli

The rear sight assembly carried the last three digits of the serial number on the base and leaf. Both parts were also marked with the inspector's initials or cartouche. The ordnance acceptance mark has not been observed on the rear sight parts.

Rear Sight Assembly Markings, Schmidt

Each part of the rear sight will show an inspector's mark and an "ordnance acceptance mark," see Figure 5-53.

The Model 1889 leaf will show a three- or four-digit fitting number plus various inspector marks on the underside. The Model 1889 rear sight base will show a variety of inspection marks between the protective ears.

Fig. 5-49. K31 barrel, intertwined "BP" acceptance mark at center.

Fig. 5-50. Last three digits of the receiver serial number on K31 barrel.

Fig. 5-51. Factory inspection (left) and Swiss military acceptance (right) markings on K31 barrel.

Fig. 5-52. Acceptance mark (arrow), inspection marks (center and far right) and subcontractor's mark (right) on K31 barrel.

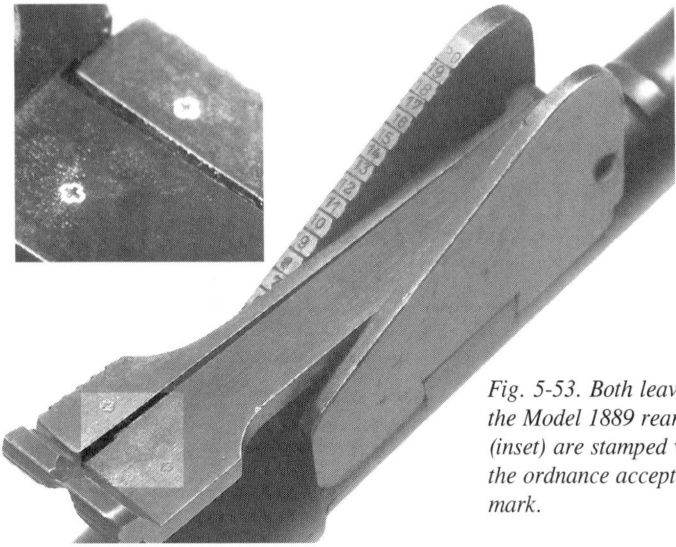

Fig. 5-53. Both leaves of the Model 1889 rear sight (inset) are stamped with the ordnance acceptance mark.

Barrel Bands

Vetterli Barrel Bands

The use and number of barrels bands on various models of the Vetterli rifles and carbines is rather complicated. The reader is referred to Table 1-2 on page 15 for a complete listing for rifles, short rifles, and carbines.

The front barrel band will be marked with an inspector's initials or cartouche below the bayonet mount and an inspector's initials on the rear face of the bayonet mount. The serial number will be marked to the left of the cleaning rod hole. A subcontractor's initials or numbers may be marked on the right side.

All barrel bands were marked with the last three digits of the receiver serial number on the front, bottom face. They were also marked with an inspector's initials or cartouche. They may also show a subcontractor's mark or lot number.

Model 1889 Schmidt Barrel Bands

The forward band was marked with the ordnance acceptance mark on the top. An inspector's number or initial will be found stamped on top of the stacking rod housing, and the stacking rod will show an ordnance acceptance mark on the bottom of the rod. The rear barrel band will be marked with the ordnance acceptance mark on the top and an inspector's initial or number on the sling swivel lug.

Model 1911 Rifle Barrel Bands

The forward band was marked with the ordnance acceptance mark on the left, lower side. The stacking rod was marked with the ordnance acceptance mark. The rear band was marked with the ordnance acceptance mark on the top.

Model 1911 Carbine Barrel Bands

The forward band was marked exactly as the forward rifle band. The rear band was marked with the ordnance acceptance mark on the bottom left side.

K31 Rifle Barrel Bands

The forward barrel band was marked with the ordnance acceptance mark on the lower right side and a contractor's mark. The rear band was marked with the ordnance acceptance mark under the side-mounted swivel lug on the left side.

Cleaning Rod

Only the Vetterli rifle and carbine were equipped with externally mounted cleaning rods. They were marked with an inspector's initials or cartouche behind the head.

Bolt Assembly Markings

Vetterli Bolt Assembly

The bolt cap was marked with an ordnance acceptance mark over an initial cartouche.

The Model 1869/1871 sheet metal main spring cover was marked only with an ordnance acceptance mark over an initial cartouche. The Model 1878/1881 main spring cover was marked with the receiver serial number and an ordnance acceptance mark over an initial cartouche.

The Model 1869/1871 bolt body was marked with the complete receiver serial number only. The Model 1878/1881 bolt body was marked with the last three serial numbers of the receiver on the left side. The inspector's cartouche, a Swiss cross over an initial, was marked on the right side.

The extractor in all models was marked on the top, rear with an ordnance acceptance mark in a cartouche.

The striker in all models was inspector marked. The firing pin in all models was marked with an ordnance acceptance mark over an initial cartouche, and occasionally an inspector's mark will also be noted.

NOTE: Bolt bodies may show a subcontractor's lot number stamped on the left side in numbers smaller than 0.01 inch.

Schmidt Bolt Assembly

All bolt parts in the Model 1889 and 1911 bolt assemblies were inspected and marked. K31 bolts were all inspected but not all parts were marked.

Model 1889 Bolt Assembly Markings

See Figure 5-54 for the markings on all the Model 1889 bolt parts.

1) The bolt body was marked with the full serial number and the ordnance acceptance mark.

2) The bolt sleeve was marked with the ordnance acceptance mark between the locking lugs. Inspector's marks in the form of letters or numbers are found on the opposite side.

3) The bolt cap was marked with the ordnance acceptance mark.

4) The firing pin was marked with the ordnance acceptance mark.

5) The cocking piece was marked on the bolt lug with the ordnance acceptance mark.

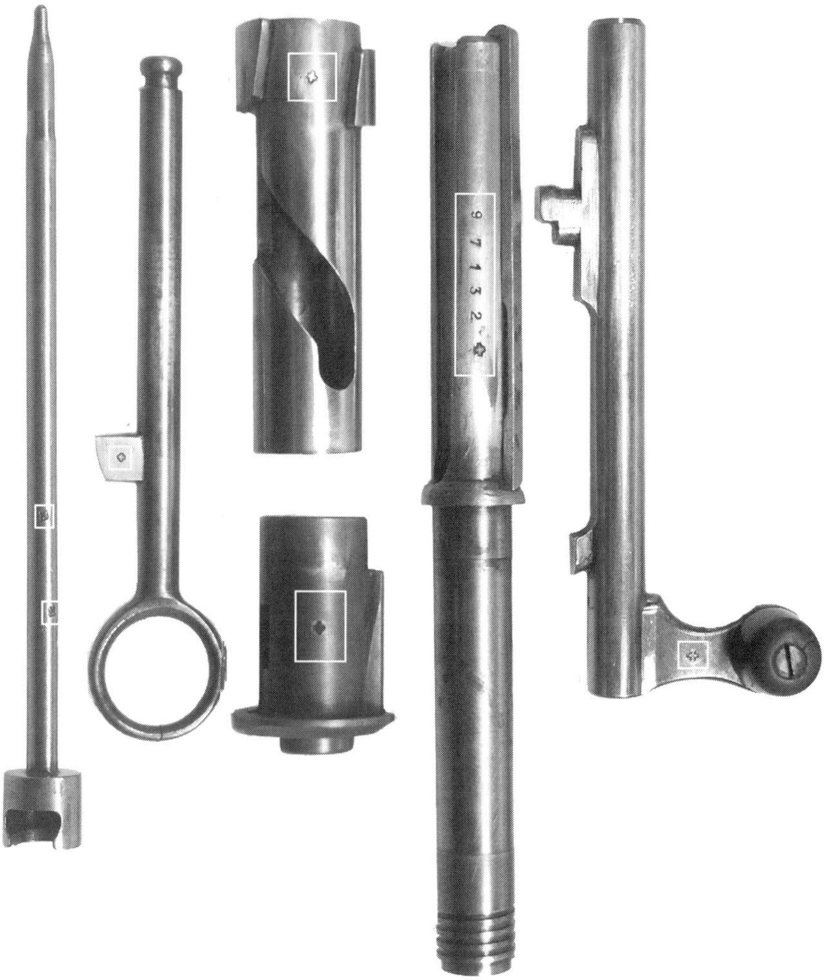

Fig. 5-54. The location of inspection and acceptance markings and serial numbers on Model 1889 bolt parts.

6) The bolt handle was marked with the ordnance acceptance mark. The bottom of the wooden bolt knob was marked with a fitting number and the last three digits of the serial number.

7) The extractor was marked on the underside with the ordnance acceptance mark.

8) The main spring was unmarked.

Model 1911 Bolt Assembly Markings

See Figure 5-55 for the location of all markings on the Model 1911 bolt parts.

1) The bolt body was marked with the full serial number and ordnance acceptance mark. Additionally, a subcontractor's mark was often found stamped on bolt body. The example in Figure 5-55 was marked WSO for the firm of Wüthrich Solothurn.

2) The bolt sleeve was marked with the ordnance acceptance mark between the locking lugs. Inspector and subcontractor marks will be found on the opposite side.

3) The bolt cap was marked on the side with the ordnance acceptance mark.

4) The firing pin was marked on the pin and the collar with the ordnance acceptance mark.

5) The cocking piece was marked with the ordnance acceptance mark and on the left side of the lug with an inspectors mark.

6) The bolt handle was marked with the ordnance acceptance mark on the upper side and an inspector's mark on the underside.

7) The extractor was marked on the underside with the ordnance acceptance mark.

8) The main spring was unmarked.

K31 Bolt Assembly Markings

See Figure 5-56 for the location of all markings on the K31 bolt parts.
1) Surprisingly enough, the several K31 bolt bodies inspected were not marked with the ordnance inspection mark or with an inspector's or subcontractor's mark.

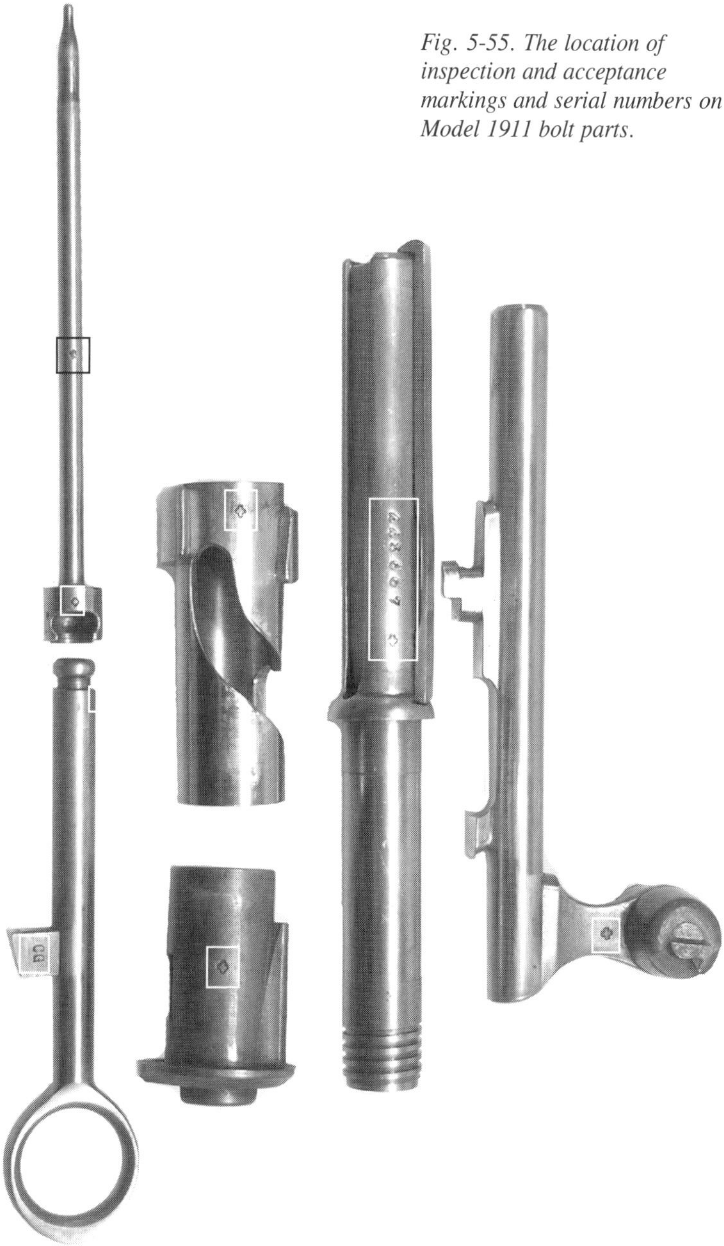

Fig. 5-55. The location of inspection and acceptance markings and serial numbers on Model 1911 bolt parts.

2) The bolt sleeve was marked with the serial number. It was also marked with the ordnance acceptance mark, and subcontractor's mark.

3) The bolt cap of several rifles examined was marked "CN" for Chrome Nickel and had the ordnance acceptance mark. It was also marked with the contractor's mark or code and followed by the last two digits of the year of manufacture, i.e., "49" for 1949.

4) The firing pin was marked with the ordnance acceptance mark.

5) The cocking piece was marked "CN" for Chrome Nickel and the ordnance acceptance mark on the right side of the lug and the contractor's mark on the left.

6) The bolt handle was marked "CV" for Chrome Vanadium and had the ordnance acceptance mark and subcontractor's marking.

7) The extractor was marked on the underside with the ordnance acceptance mark, inspector's, as well as the subcontractor's markings.

8) The main spring was unmarked.

Finishes

All wood parts were sanded smooth and oiled, possibly by dipping in a heated oil bath. Stock wood was European walnut (*Nussbaumholz*) and, after 1946 (at circa serial #868,901), beech *(Buchenholz)*. European walnut has a lighter hue than American black walnut and will range in color from dark coffee to light yellow when oiled. Beech is a yellowish wood with a distinctive layered look to the grain. It takes on a medium caramel color when oiled.

If the stock is exceptionally dirty, it can be washed in a mild detergent solution using a soft-bristle brush. This treatment will not remove the original oils. For caked-on oil and grime, use one of the new organic, citrus-based cleaning solutions rather than one based on mineral spirits, alcohol or acetone. You will achieve the same results without exposing yourself to carcinogenic organic solvents.

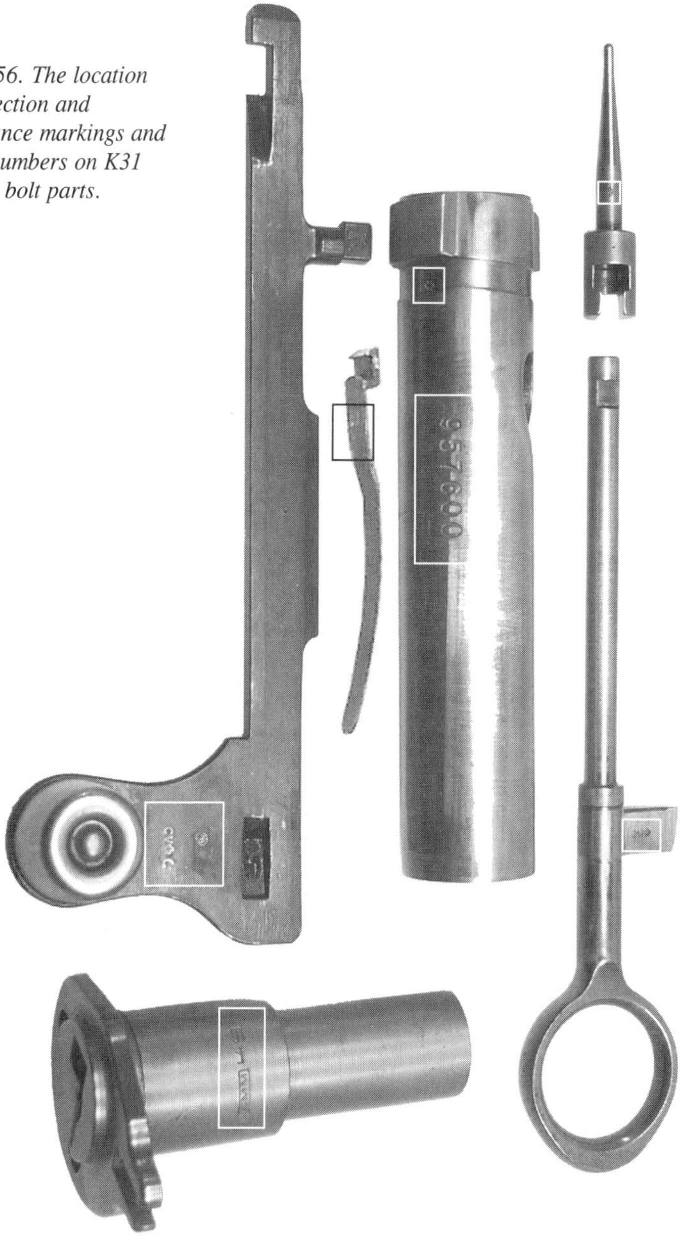

Fig. 5-56. The location of inspection and acceptance markings and serial numbers on K31 marked bolt parts.

Steel parts, except for receivers, were blued using a hot salt bath method after polishing and degreasing. Receivers were heat-treated in special ovens. The heat treatment hardened the steel and produced a tough surface similar to case-hardening. It also gave the receiver its deep blue-black color.

Vetterli Parts and Finishes

Screws and small parts were blued by heating. Springs were colored in the tempering process. The Vetterli bolt cap, spring cover and handle were blued; all other bolt parts were polished and left in the white. Magazine tubes were left in the white. Butt plates, receivers, barrels, barrel bands, sights and trigger guard assemblies were blued. The cartridge lifter and lifter assembly were left in the white.

Schmidt Parts and Finishes

All Schmidt bolt parts were left in the white, except for the K31 bolt operating rod (also called a cam follower rod) which was blued. The bolt cylinder, bolt locking sleeve, cocking piece, bolt operating rod for all but the K31 were polished bright after finishing and heat treatment. Bolt caps were left in the white but were not polished. Receivers, butt plates, barrels, barrel bands, sights, sling swivel and plates, trigger assemblies and magazines and their parts were blued. The collector should note that the underside of the upper tang was polished and left in the white. During refurbishment, if the receiver was reblued, this part was rarely repolished and is a good indicator as to whether or not the receiver has been reblued. Refer to Figure 5-29. Do not mistake natural aging for an aged blue finish.

The markings on steel parts that have been reblued will be reduced in depth and it will appear as if their top edges have been sheared off. Sharp metal edges will have been rounded during polishing. Receivers that have been refinished will not show a polished surface at the rear underside of the receiver where the trigger bearing presses against the receiver.

Chapter 6
Bayonets

Like the Vetterli and Schmidt rifles and carbines, their bayonets are well designed and of high quality, see Figure 6-1. Three bayonet models were manufactured for the Vetterli rifles and three more for the Schmidt rifles and carbines. Each model had several variations.

Fig. 6-1. Model 1911 rifle with Model 1889/99 knife bayonet. North Cape Publications collection.

Model 1871 Socket Bayonet *(Stichbajonett)*

The first bayonet manufactured for the Model 1869/71 Vetterli rifle was a variation of the Model 1863 socket bayonet developed for the Swiss Peabody rifle. The Model 1871 had a quadrangular (four-edged) blade and was cruciform in cross section. The socket had an "L"-shaped cut that allowed it to be fitted over the rifle muzzle, then turned clockwise to clear the front sight. A locking ring fitted with a tension screw was

Fig. 6-2. Model 1871 socket bayonet for the Vetterli Rifle. North Cape Publications collection.

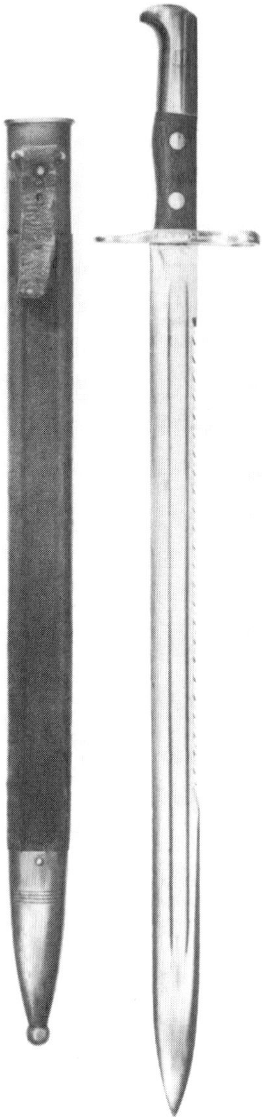

Fig. 6-3. Model 1878 sword bayonet with saw back for the Vetterli rifle. North Cape Publications collection.

tightened down behind the front sight base to lock the bayonet in place, see Figure 6-2.

Model 1878 Sword Bayonet with Saw Back (*Säbelbajonett*)

The 1871 angular bayonet was replaced by Act of Parliament on April 30, 1878, with the Model 1878 saber bayonet with a saw-edged back, see Figure 6-3. The Model 1878 bayonet was designed for the Model 1878 Vetterli rifle and used on the Model 1881 as well. The saw back enabled the soldier to cut the wood stakes on which barbed wire was stapled or bundles of reeds and branches to bridge ditches, clear brush and build field shelters. For efficient cutting, the saw teeth were angled forward, see Figure 6-4. The saw teeth were set far enough back on the blade that it still made an efficient bayonet.

The Model 1878 bayonet was 23.5 inches long overall and had a straight "knife" blade 18.6 inches long and 1.1 inches wide. The back was 0.122 inch thick at the guard. The bayonet was ground to a semi-spear point and had a 3-inch-long false edge. The saw back was 10.2 inches long with 25 teeth on the right side and 24 on the left. The ricasso was plain and merged into a fuller on the obverse side only which was 14.2 inches long. The reverse side lacked a fuller.

The hilt was 4.45 inches long. Two hard leather grip panels 2.67 inches long were fastened to the hilt by four rivets. The bayonet catch, or stud lock, was a spring-loaded stud. It was mounted on the right side of the hilt; the triangular stud projected into the bayonet catch channel, or stud guide, cut into the top of the hilt, see Figure 6-5. The spring was secured by a screw in the right side of the hilt and was 2.7 inches long.

Fig. 6-4. The Model 1878 bayonet has 25 teeth on the right and 24 on the left.

The guard was 3.8 inches long and the top branch was pierced for the rifle muzzle. The bore was 0.7 inch in diameter.

The scabbard was 19.6 inches long, 1.5 inches wide and made of leather dyed black. The throat and tip were of blued iron. The tip was 4.2 inches long and ended in a ball tip. It had four horizontal lines scribed across its front 0.9 inch from the top. The tip was secured to the leather body by a single rivet on the obverse side and two on the reverse, refer to Figure 6-3.

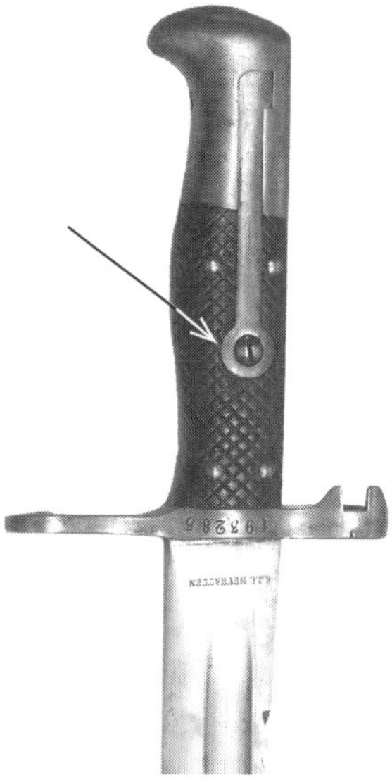

Fig. 6-5. Model 1878 bayonet hilt. The bayonet spring catch is inletted into the right side (arrow).

The throat was 2.35 inches long. An iron staple was brazed or welded to the throat. The throat was held to the leather body by two rivets on either side. The scabbard was held to the soldier's belt with a leather frog. The staple protruded through the frog. A strip of leather riveted to the staple fit through a buckle on the frog to secure the scabbard to the frog. All metal parts of the scabbard were blued.

The bayonet blade was polished bright and marked with the manufacturer's name and city on the obverse side of the ricasso. The reverse side was blank. Inspector's marks are found on top of the hilt near the bayonet stud channel. The serial number of the rifle for which the bayonet was issued was usually stamped onto the right side of the guard in line with the manufacturer's name on the ricasso. The scabbard inspector's mark or cartouche is usually found on the obverse side of the ball tip. A serial number or other marking may be found on the obverse side of the throat.

A variation of the Model 1878 saber bayonet with saw back is identical with the exception of a ball-shaped stud inserted into the fuller about three inches from the guard. The stud was used to secure the bayonet in the scabbard and prevent loss or rattling. It is not known to the author if the bayonets were manufactured this way or were later refitted with the stud.

Model 1887 Sword Bayonet with Saw Back *(Säbelbajonett)*

The Model 1887 was developed for use with the Model 1878 and 1881 Vetterli rifles. The dimensions and overall form are quite similar to those for the Model 1878 sword bayonet with saw back. The exceptions are 1) a slightly wider fuller on *both* sides of the blade, 2) a greater number of saw teeth to increase the bayonet's efficiency: 33 on the right and 32 on the left, 3) three rivets to secure the leather guard panels to the hilt and 4) a milled fitting on the right side of the guard to fit around the front sight base.

At least two variations of the Model 1887 sword bayonet have been observed. One variation is a plain sword bayonet without the saw back. Another is a saw back with 50 teeth on the right and 49 on the left.

Model 1889 Schmidt Bayonet (*Dolchbajonett*)

The Model 1889 knife bayonet is 16.6 inches long overall with a blade 11.75 inches long and 1 inch wide and 0.2 inch thick at the guard, see Figure 6-6. The hilt is 4.7 inches long and has two (walnut) wooden grip scales 2.7 inches long fastened with two rivets ground flat. The bayonet stud is of the Mauser type with the spring mechanism inside the hilt. The bayonet stud channel is 1.4 inches long, see Figure 6-7.

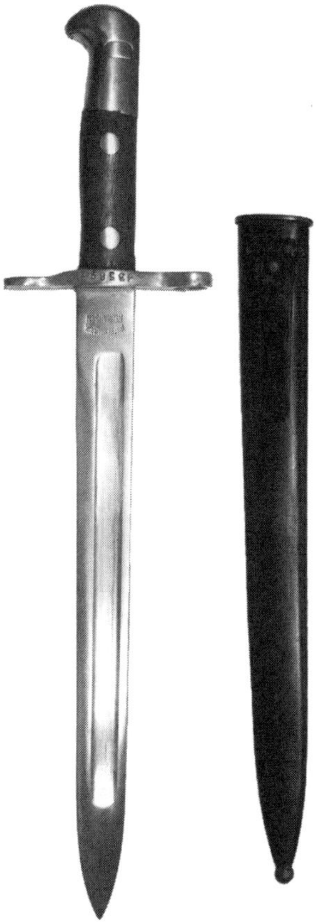

Fig. 6-6. Model 1889 bayonet and scabbard for the Schmidt rifle. North Cape Publications collection.

The blade has a spear point and a false edge 1.2 inches long. Fullers are ground into each side of the blade and are 8.2 inches long. The ricasso is plain and carries the name of the manufacturer and city on the obverse. The reverse is blank.

The cross guard is 3.7 inches long and the top of the guard is pierced for the rifle muzzle. It is 0.6 inch in diameter. The serial number of the rifle for which the bayonet was issued is often stamped on the right side, center of the guard, see Figure 6-8. If the bayonet was reissued for a different rifle, the old serial number was ground off and the new serial number stamped on. It is often possible to see the remains of the previous serial number.

The bayonet blade, guard and hilt were polished bright.

The scabbard was made of iron and was 12.5 inches long. The scabbard ended in a ball tip. The throat with spring clamps was

Fig. 6-7. A Mauser-style bayonet catch mechanism was used on the Model 1889 bayonet.

applied separately and secured to the body with a single screw on the reverse side. A metal staple was brazed or welded to the body, and a leather strip or tongue was fastened to the staple with a single brass or copper rivet. A frog was worn on the soldier's belt. The scabbard slid into the frog, and the leather tongue was threaded through a buckle on the frog to secure it.

Early scabbards were brazed together; later scabbards were welded. The seam can usually be detected on the right side of the scabbard. Look just to the right of the staple.

An inspector's mark or symbol will be found on the bayonet at the start of the bayonet stud channel. No inspection markings of any kind have been detected on the scabbards.

Fig. 6-8. The serial number of the rifle or carbine for which the bayonet was issued will often, but not always, be found stamped on the guard.

NOTE: The Model 1889 Schmidt bayonet served as the model for the original Model 1892 Krag bayonet used by the regular U.S. military from 1892 to 1903.

Model 1889/92 Schmidt Bayonet (*Stichbajonett*)

These bayonets were designed and manufactured as a cost-savings feature. They were made from the Model 1871 angular bayonet and were all steel, see Figure 6-9. A new guard was fabricated and brazed or welded onto the cruciform blade. The end of the bayonet blade was reshaped to provide a mounting for the bayonet locking stud. This is a fairly rare bayonet and was intended for use on the Model 1889 series of rifles. The bayonet was blued.

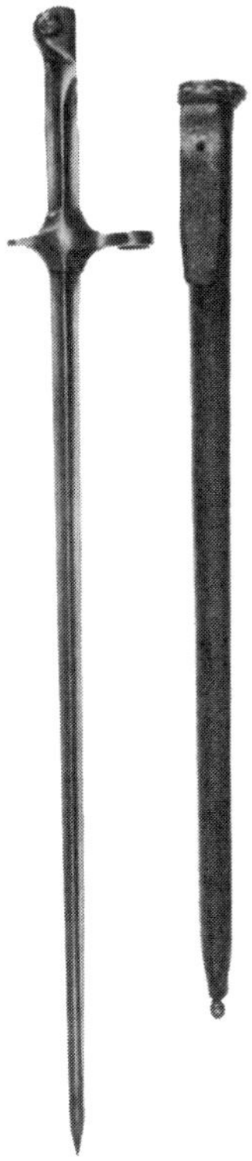

Fig. 6-9. The Model 1889/92 bayonet was remanufactured using obsolete Model 1871 angular bayonets. North Cape Publications collection.

The scabbard was made of leather and designed to fit into a variation of the Model 1889 bayonet frog.

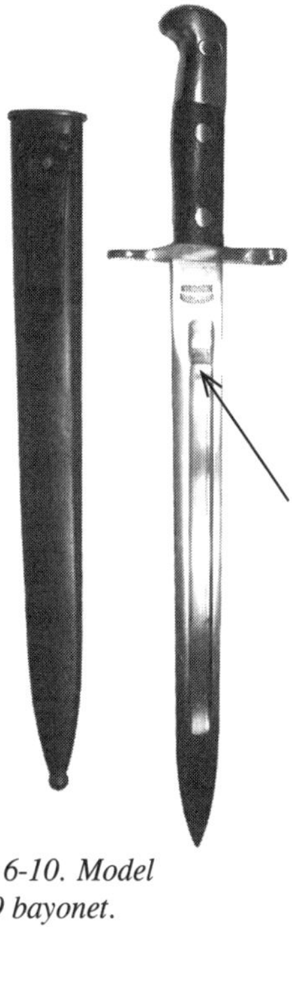

Fig. 6-10. Model 1899 bayonet.

Model 1899 Schmidt Bayonet (*Dolchbajonett*)

The Model 1899 Schmidt bayonet and scabbard are identical to the Model 1889 with the single exception of the fuller. A sloping ridge was left 2 inches below the guard on either side when the fuller was ground, see Figure 6-10, arrow. The ridge served to grip the throat springs more tightly and prevent the bayonet from falling out of, or rattling in, the scabbard.

Model 1908 Schmidt Bayonet (*Stichbajonett*)

Very similar to the Model 1889/92 bayonet. The scabbard was made of iron and blued.

Model 1911 Schmidt Bayonet with Saw Back (*Säbelbajonett*)

The Model 1911 bayonet was designed to be used with the Model 1911 carbine only because of its smaller bore. It was issued to Pioneer, and to infantry troops, equipped with the carbine. The bayonet was an updated version of the Model 1887 bayonet, see Figure 6-11.

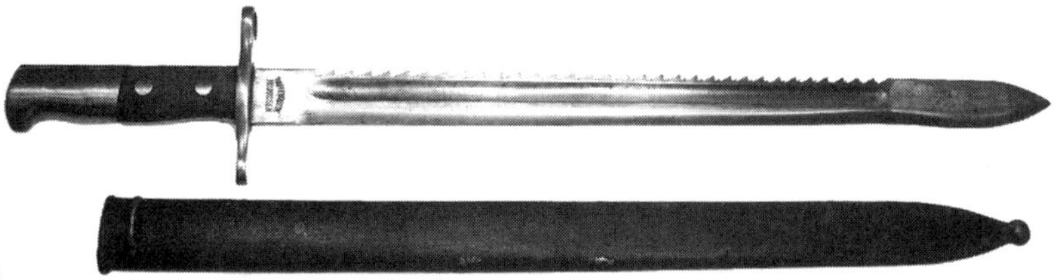

Fig. 6-11. The Model 1911 saber bayonet with the saw back was issued with the M1911 carbine to Pioneer and later infantry troops. North Cape Publications collection.

It was 24 inches long overall with a blade 18.75 inches. The blade was 1.1 inches wide at the guard and the same at the widest part of the tip which swelled slightly wider than the center part of the blade. The saw-toothed section is 13.2 inches long and begins 1.3 inches below the guard. The blade has 32 teeth on the right and 33 on the left. The hilt is 4.8 inches long and has two walnut wood scales 2.4 inches long. They are held to the hilt with two rivets. The bayonet catch mechanism is exactly the same as in the Model 1889 series for the Schmidt rifle. The scabbard is made of iron and is 20 inches long. It ends with a ball tip. The scabbard is an elongated version of the Model 1889 series.

NOTE: A variation of the Model 1911 Schmidt bayonet with saw back was made using the Model 1878 and 1887 blades mated with the Model 1889 hilt. They can easily be distinguished as the blades are straight spear points and do not swell behind the point

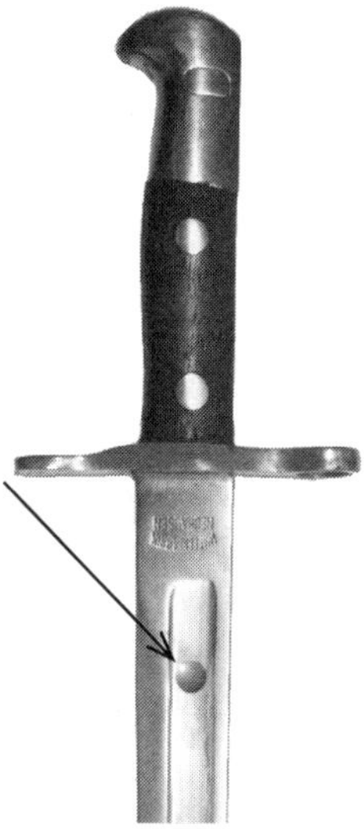

Fig. 6-12. Model 1889/18. Note the hemispherical stud (arrow). North Cape Publications collection.

Model 1889/1918 Schmidt Bayonet (*Dolchbajonett*)

The Model 1889/1918 bayonet and scabbard are identical to the Model 1899 bayonets but with a hemispherical stud welded to the blade, see Figure 6-12, to place greater pressure on the throat springs and keep the bayonet from falling out of or rattling in the scabbard. These bayonets are the Model 1889 with the stud added.

Model 1918 Schmidt Bayonet (*Dolchbajonett*)

The Model 1918 bayonet was similar to the Models of 1889, 1899 and 1889/18 with the exception of a hilt 0.1 inch longer, and a double-edged blade with a cruciform, reversed fuller, see Figure 6-13. It was developed for use with the Model 1911 carbine and the K31. It was also issued with the AK53 Rifle. The scabbard was the Model 1889.

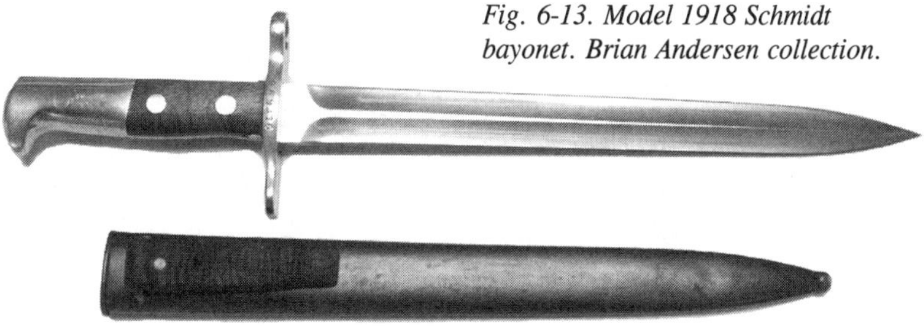

Fig. 6-13. Model 1918 Schmidt bayonet. Brian Andersen collection.

Appendices

Appendix A
Exploded Views

It is remarkable that a pioneering design such as the Vetterli underwent so few changes during its 21-year service life. Although four models of rifles, three short rifles, seven carbines and two cadets were built, the changes to the functioning mechanism were minor.

The Schmidt rifle and carbine, on the other hand, progressed through three major model changes during its 69 years of service. And between the first, the Model 1889 and the second, the Model 1911, three variations were designed, produced and fielded to the Swiss military. They were the Models of 1889/1896, 1889/1900 and 1905.

The three variations of the Model 1889 were unique production runs; that is to say they were not modifications of previous versions. The Model 1889 with the changes made in 1896 was produced as a "new rifle." Previous Model 1889 rifles were not rebuilt. And so with the Model 1889/1900 and the Model 1905.

The Model 1911 was produced in only two variations, the infantry rifle and the carbine. The K31 was produced in two variations, the standard issue military rifle (K31) and the ZFKar.31/55 Sniper Rifle. The ZFKar. 31/55 sniper rifle was manufactured specifically as a sniper rifle with the receiver canted to permit a scope to be mounted above the receiver and in line with the bore.Two other sniper versions of the K31 were developed during World War II, the ZFKar. 31/42 and ZFKar. 31/43, but these were essentially K31 rifles equipped with receiver-mounted telescopic sights.

The exploded view overleaf illustrates the parts of the Schmidt rifle as well as those for successive models that differ significantly in appearance from the parts designed for the Model 1889. Not every single part for every model and variation could be shown for reasons of space. Instead, the reader is referred to Chapter 3 for a complete discussion of parts and their changes by model and variation.

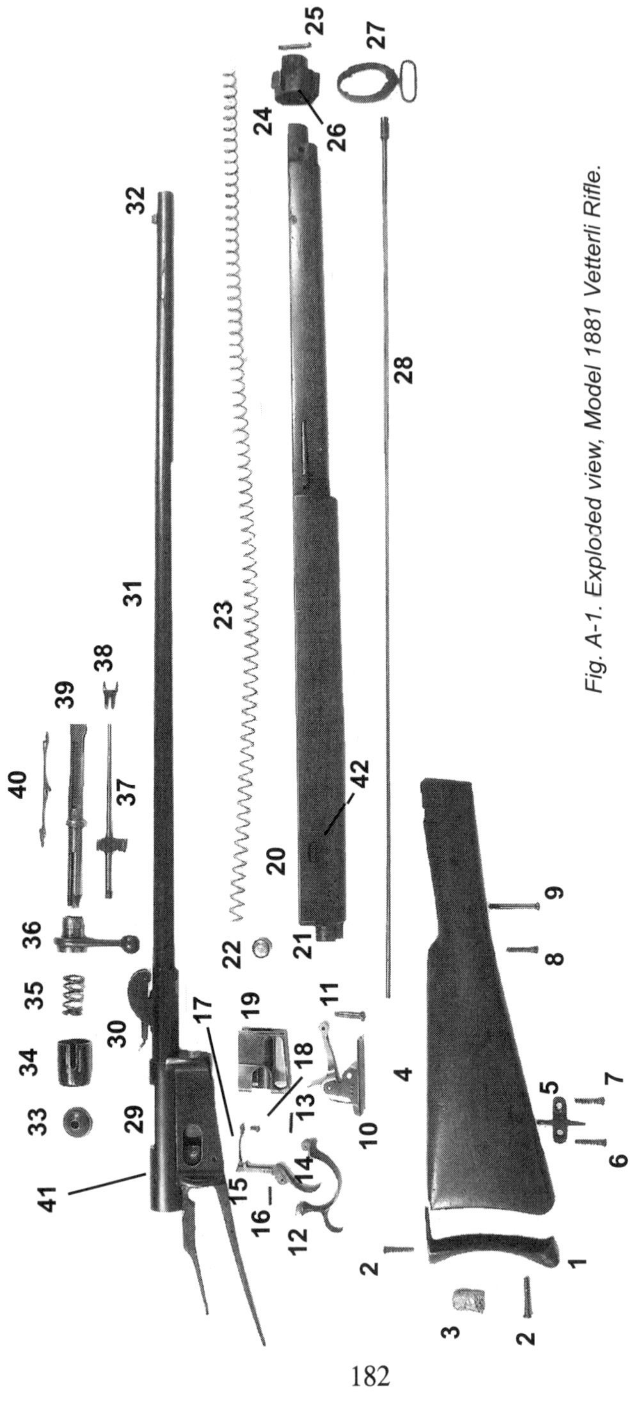

Fig. A-1. Exploded view, Model 1881 Vetterli Rifle.

Vetterli Rifle, Model 1881 Exploded View

1. Butt Plate
2. Butt Plate Screws
3. Spare Firing Pin (wrapped in tin foil)
4. Buttstock
5. Rear Sling Swivel
6. Rear Sling Swivel Screw
7. Rear Sling Swivel Screw
8. Trigger Guard Screw
9. Tang Screw
10. Cartridge Lifter Assembly
11. Cartridge Lifter Screw
12. Trigger Guard
13. Trigger Guard Pin
14. Trigger
15. Sear
16. Trigger/Sear Pin
17. Sear Spring
18. Sear Spring Screw
19. Elevator
20. Forend
21. Magazine Tube
22. Magazine Follower and End Cap
23. Magazine Spring
24. Forward Barrel Band
25. Forward Barrel Band Screw
26. Bayonet Mount
27. Rear Barrel and Sling Swivel
28. Cleaning Rod
29. Receiver
30. Rear Sight
31. Barrel
32. Front Sight
33. Bolt Cap
34. Main Spring Cover
35. Main Spring
36. Bolt Handle with Locking Lugs
37. Striker with Cocking Lug
38. Rimfire Firing Pin
39. Bolt Body
40. Extractor
41. Bolt Retainer Key
42. Forend Key

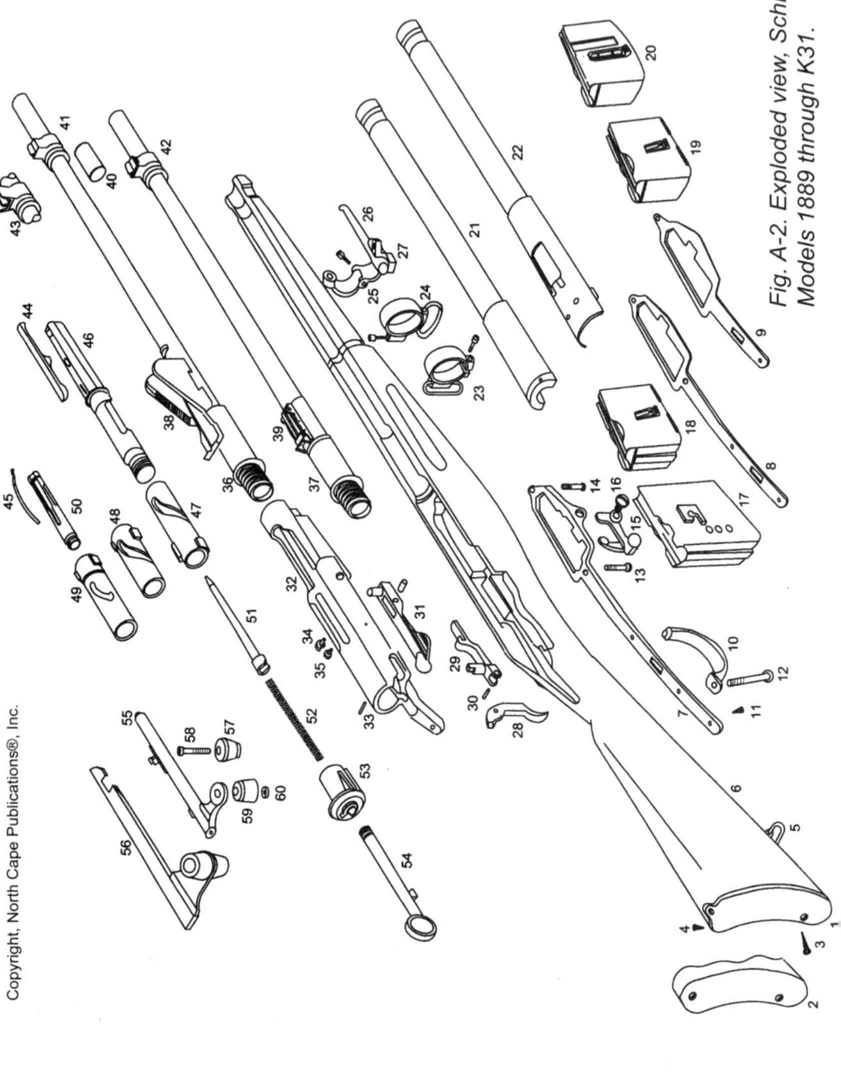

Fig. A-2. Exploded view, Schmidt Rifles, Models 1889 through K31.

Schmidt Rifle and Carbine, Exploded View

1. Butt Plate, Model 1889
2 Butt Plate, Model 1911, K31
3. Lower Butt Plate Screw
4. Upper Butt Plate Screw
5. Rear Sling Swivel
6. Stock
7. Trigger Guard Plate, Model 1889
8. Trigger Guard Plate, Model 1911
9. Trigger Guard Plate, K31
10. Trigger Guard
11. Rear Trigger Guard (wood) Screw, Model 1889
12. Tang Screw
13. Trigger Guard Screw, Model 1889, 1911
14. Trigger Guard Screw
15. Magazine Cutoff, Model 1889
16. Magazine Cutoff Screw, Model 1889
17. Magazine, Model 1889
18. Magazine, Model 1889/96, 1889/00, 1905
19. Magazine, Model 1911
20. Magazine, K31
21. Hand Guard, Rifle, Model 1889, 1911, K31
22. Hand Guard, Carbine, Model 1911
23. Barrel Band, Rear and Screw, Model 1911 Carbine
24. Barrel Band, Rifle, Rear and Clamp Screw
25. Barrel Band, Front and Screw
26. Stacking Rod
27. Bayonet Mount
28. Trigger
29. Sear
30. Trigger Pin
31. Bolt Release Assembly
32. Receiver
33. Sear Pin
34. Ejector
35. Ejector Lock Screw
36. Barrel, Model 1889
37. Barrel, Model 1911, K31
38. Rear Sight, Model 1889
39. Rear Sight, Model 1911, K31
40. Barrel Collar
41. Front Sight, Rifle, Model 1889
42. Front Sight, Rifle, Model 1911
43. Front Sight, Model 1911 Carbine, K31
44. Extractor, Model 1889, 1911
45. Extractor, K31
46. Bolt Cylinder, Model 1889, 1911
47. Bolt Locking Sleeve, Model 1889
48. Bolt Locking Sleeve, Model 1911
49. Bolt Locking Sleeve, K31
50. Bolt Cylinder, K31
51. Firing Pin
52. Bolt Spring
53. Bolt Cap
54. Cocking Piece/Striker
55. Bolt Handle, Model 1889, 1911
56. Bolt Handle, K31
57. Bolt Knob, Upper
58. Bolt Knob Screw
59. Bolt Knob, Lower
60. Bolt Knob Nut

Appendix B
Disassembly and Reassembly

The various models of the Vetterli rifle and carbine are sufficiently alike that only one set of instructions is required.

1. Begin by opening and drawing back the bolt to assure that the breech and magazine are empty, see Figure B-1.

2. Remove the bolt by pressing out the bolt retaining key with the tip of a screwdriver. Place the screwdriver tip on the grooved right side of the key and gently tap out from right to left. The key will not come all the way out, see Figure B-2.

3. With the proper size screwdriver blade, unscrew the large screw on the left rear side of the receiver that holds the lifter assembly and remove, see Figure B-3.

4. Remove the screws that are holding the trigger guard plate, see Figure B-4.

5. Remove the lifter assembly, see Figure B-5. The lifter assembly is attached to the trigger guard and can be removed without taking out the trigger guard plate in the Models of 1869 and 1871. It is on a separate plate in the Models of 1878 and 1881.

6. The cartridge lifter can be removed through the bottom of the receiver, see Figure B-6.

7. The bolt can now be removed from the receiver, see Figure B-7.

8. Remove the stock by pulling it backward, see Figure B-8.

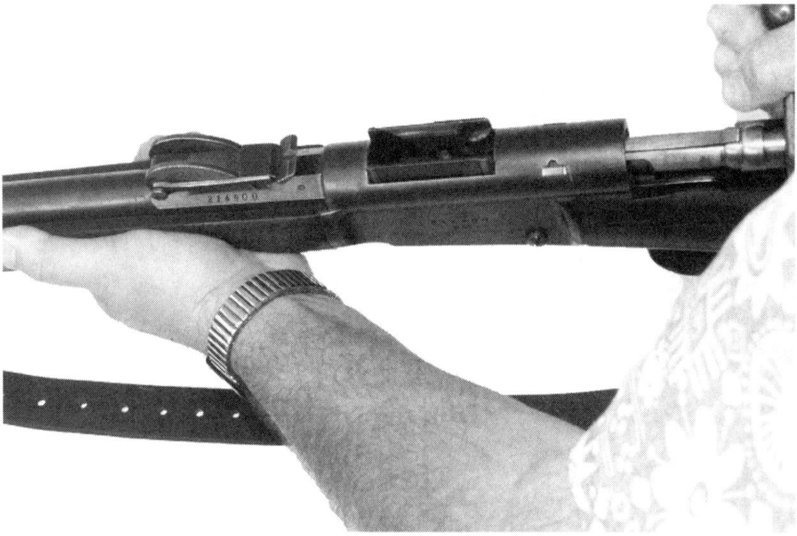

Fig. B-1. Check to make certain the chamber is empty.

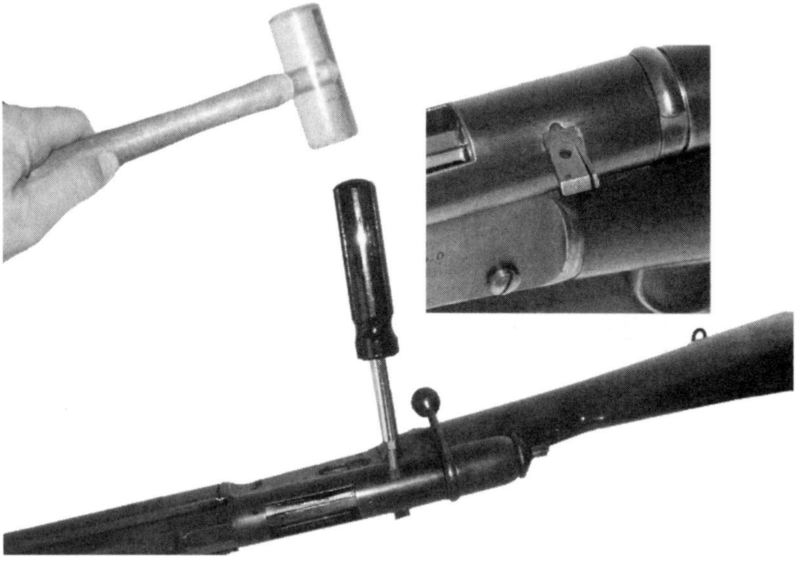

Fig. B-2. Drive out the bolt retaining key.

Fig. B-3. Remove the lifter assembly screw.

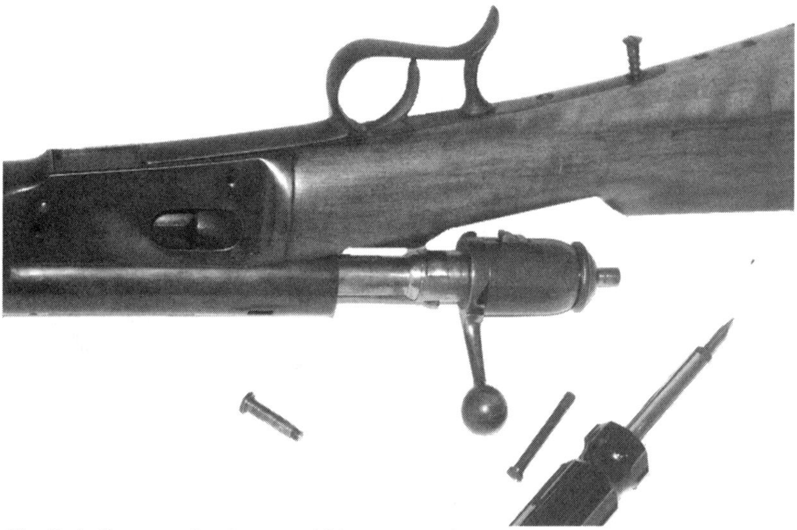

Fig. B-4. Remove the tang and trigger guard screws.

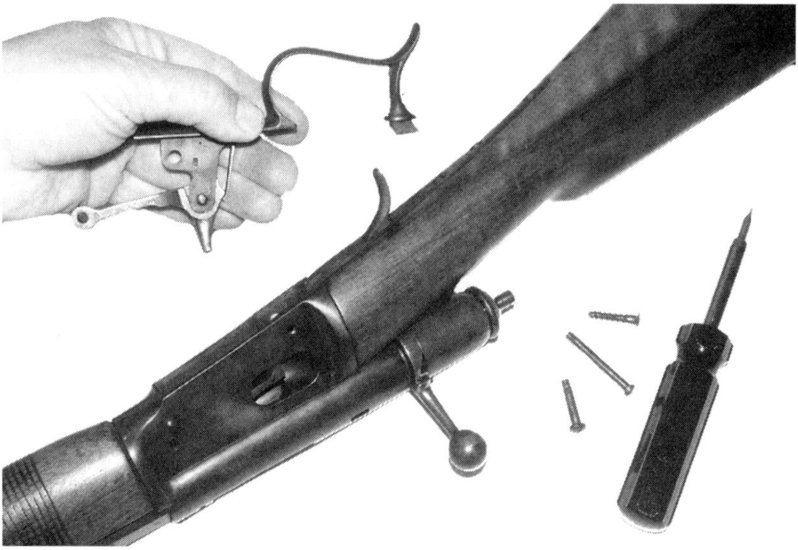

Fig. B-5. Remove the trigger guard.

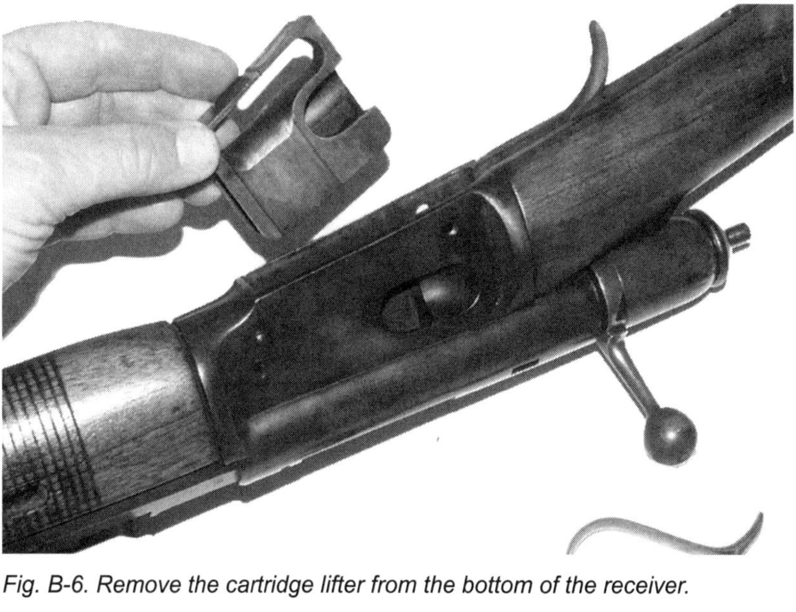

Fig. B-6. Remove the cartridge lifter from the bottom of the receiver.

Fig. B-7. Remove the bolt.

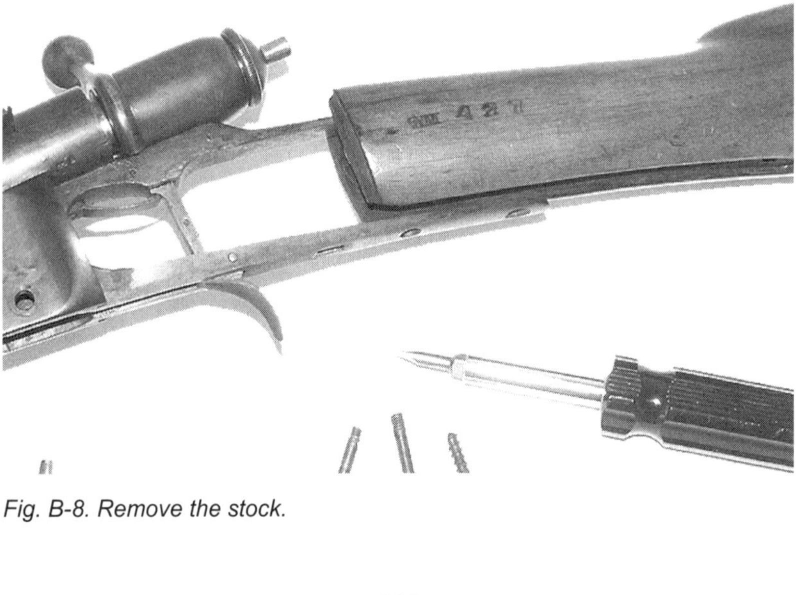

Fig. B-8. Remove the stock.

9. Remove the cleaning rod.

10. Remove the screw holding the forward barrel band and draw it off the end of the stock, see Figure B-9.

11. The trigger/sear assembly can be removed from the receiver by driving out the trigger pin from the right with the proper size punch.

12. Do not try to remove the trigger guard plate.

13. Loosen the clamp screw on the middle (Model 1869 only) and rear barrel bands and draw off the end of the stock, see Figure B-10.

14. Drive out the key holding the forend (rifle) or forearm (carbine) to the barrel. Note that the Model 1869 rifle does not have a key. Remove the forend or forearm by drawing it slightly ahead and down.

15. Unscrew the knurled ring holding the magazine follower in place and remove the magazine tube, spring, follower and rod guide.

16. Remove the butt plate screws and remove the butt plate. If the butt plate does not come off easily, treat the junction of metal and wood with lemon oil, allow to soak for an hour, then tap gently with a plastic or rubber hammer. Do not use excessive force or you will splinter the wood, see Figure B-11.

17. To disassemble the bolt, unscrew the bolt cap. Hold the cap securely as the spring may be under tension, see Figure B-12. (Wear safety glasses.)

18. Remove the bolt spring and draw out the striker, see Figure B-13.

19. Pull the bolt handle to the rear to remove, see Figure B-14.

20. Tip the bolt to the side and slide out the double-pronged rimfire firing pin, see Figure B-15.

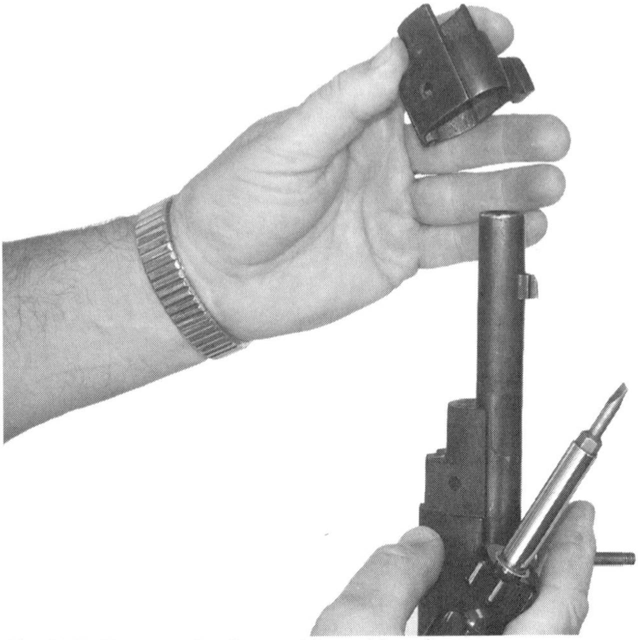

Fig. B-9. Remove the forward barrel band.

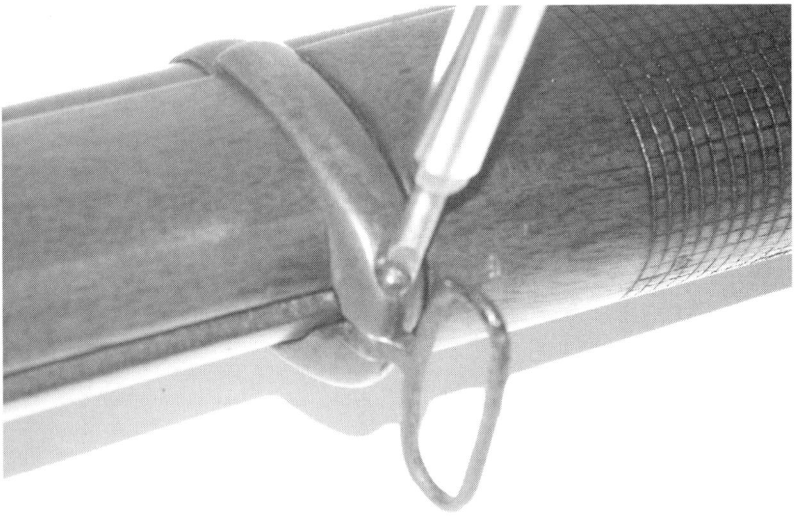

Fig. B-10. Loosen the clamping screw and remove the rear barrel band.

Fig. B-11. Remove the butt plate screws.

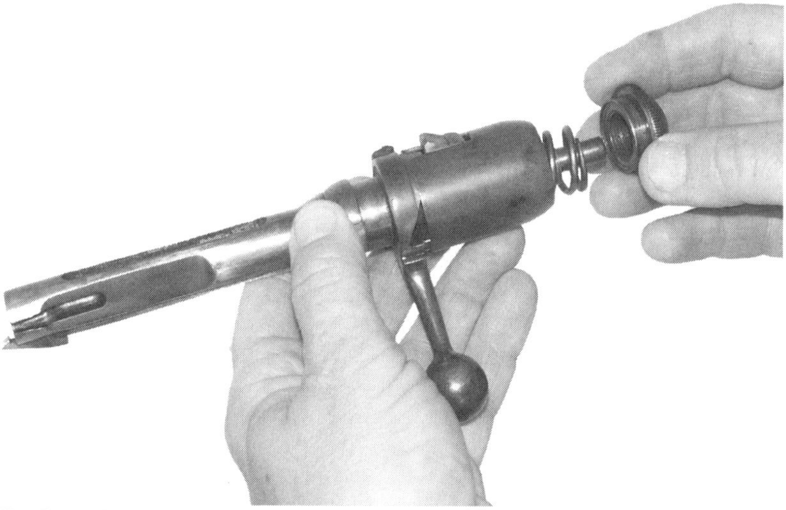

Fig. B-12. Remove the bolt cap from the bolt. Be careful as the spring may be under tension. Safety glasses should be worn for this operation.

Fig. B-13. Remove the striker/cocking piece.

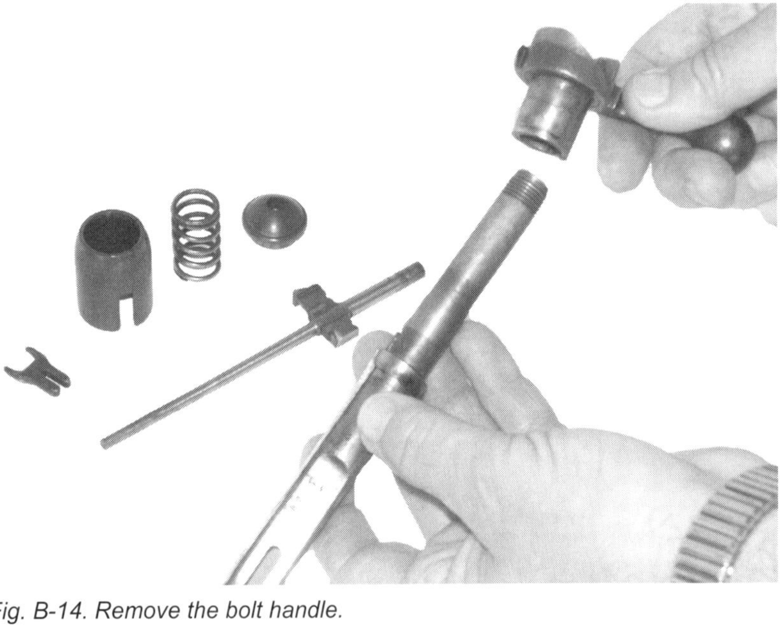

Fig. B-14. Remove the bolt handle.

21. To remove the extractor, push straight back on the tip with a screwdriver blade, or press the tip of the extractor against a hard surface and push or tap the bolt forward. Lift off. See Figure B-16.

22. There are two ways to replace the extractor in the bolt body.

A) Seat the extractor in the groove on top of the bolt body so that the hook on the underside of the extractor is behind the pin in the groove. Press down on the center of the extractor and push forward.

B) If method A fails, seat the extractor in the groove as above with the hook slightly behind the pin. Assemble the bolt handle to the bolt body. While holding the extractor down with your thumb, push the bolt handle forward until both the extractor and bolt handle are seated.

23. When reassembling the bolt, the bolt handle should be at approximately 120 degrees below and to the right of the extractor.

24. Insert the double-pronged rimfire firing pin into the slot in the bolt body with its prongs forward, followed by the striker with the cocking stud down so that it slides into the V-shaped cut in the bolt handle root.

25. Add the spring, sheet metal bolt cover and attach the bolt cap nut.

26. Reinstall the magazine components and attach the forend or forearm and stock to the barreled action.

26. Insert the bolt into the receiver but do not push the bolt key home.Assemble the cartridge lifter to the lifter assembly with the lifter arm up.

27. Insert the lifter assembly into the bottom of the receiver, first hooking the stud on the rear of the trigger guard into the rectangular hole in the trigger guard plate (Model 1871 rifle or carbine only).

28. Insert into the left side of the receiver and tighten the screw to hold the lifter assembly in place. Push home the bolt retaining key.

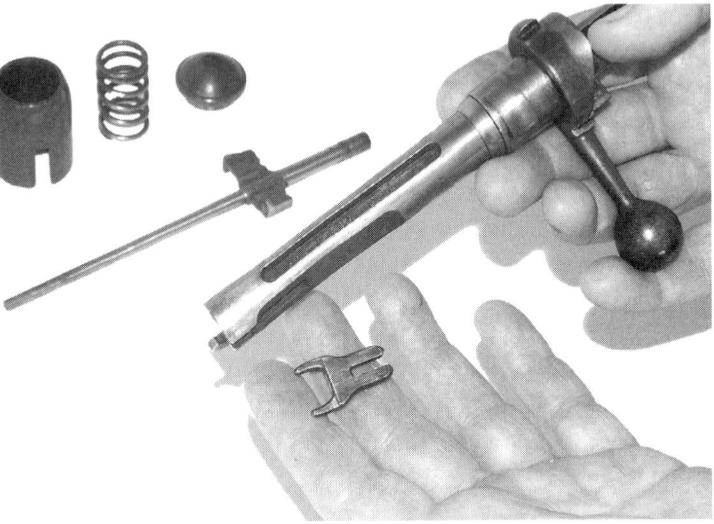

Fig. B-15. Remove the Vetterli's two-prong rimfire firing pin.

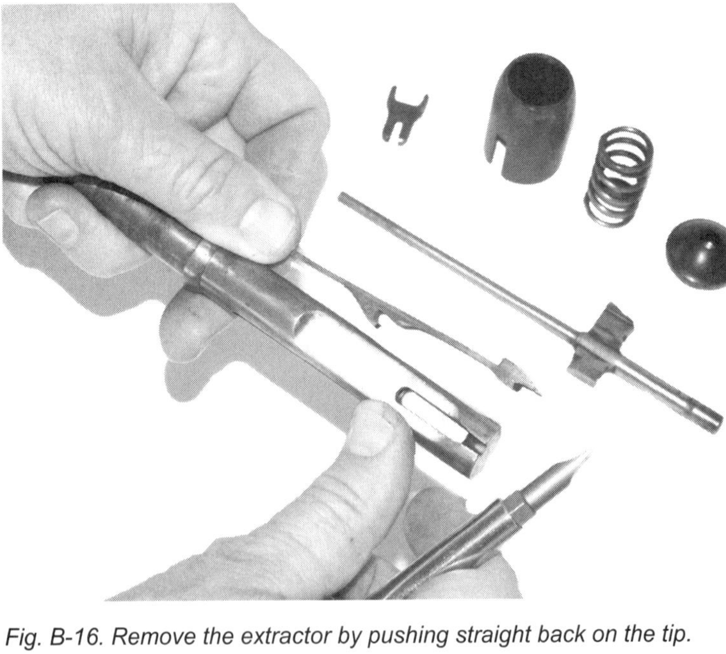

Fig. B-16. Remove the extractor by pushing straight back on the tip.

Models of 1889, 1911 and K31
Disassembly and Reassembly

The three models of the Swiss bolt action, box magazine fed rifles are sufficiently alike that a generalized description of their disassembly and assembly should suffice.

An excellent tool for dismounting any of the Swiss rifles is, of course, a Swiss Army knife, see Figure B-17. The first Swiss Army knives were manufactured by the company that is now Victorinox and issued to Swiss military personnel in 1891. Today, they remain an item of issue equipment to every Swiss soldier.

Before doing anything else, draw back the bolt and look carefully into the chamber and the magazine to make certain that no cartridges are present, see Figure B-18.

1. Pull back on the cocking ring and make a 1/8th turn clockwise so that the cocking lug rests on the rear of the bolt cap between the two rectangular slots, see Figure B-19.

2. Remove the bolt by pressing down on the bolt release, see Figure B-20.

3. Remove the magazines from the Models of 1889/96 through the K31. The Model 1889 magazine was not meant to be removed without first removing the trigger guard assembly, see Figure B-21.

4. Loosen the clamping screw on the rear barrel band, see Figure B-22.

5. Loosen the clamping screw on the front barrel band, see Figure B-23.

6. Remove the front barrel band by lifting the top of the band all the way to the left on its hinge and sliding the band off by pushing down.

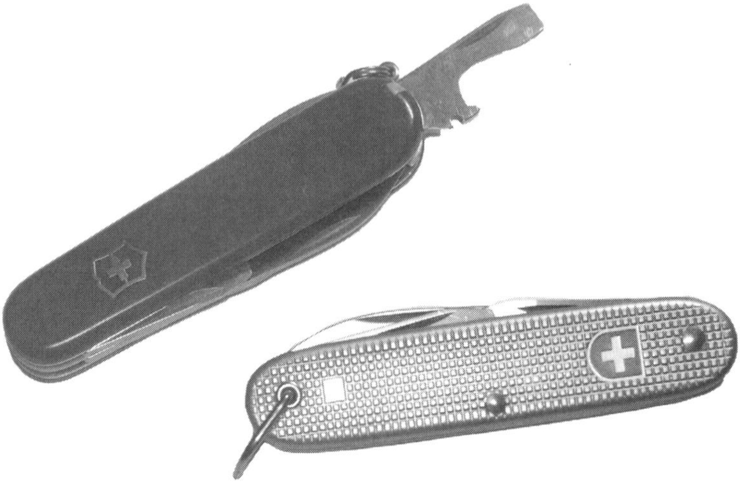

Fig. B-17. Top, Victorinox Swiss Army Knife, Tinker commercial model; below, Soldier model which is the current Swiss military issue.

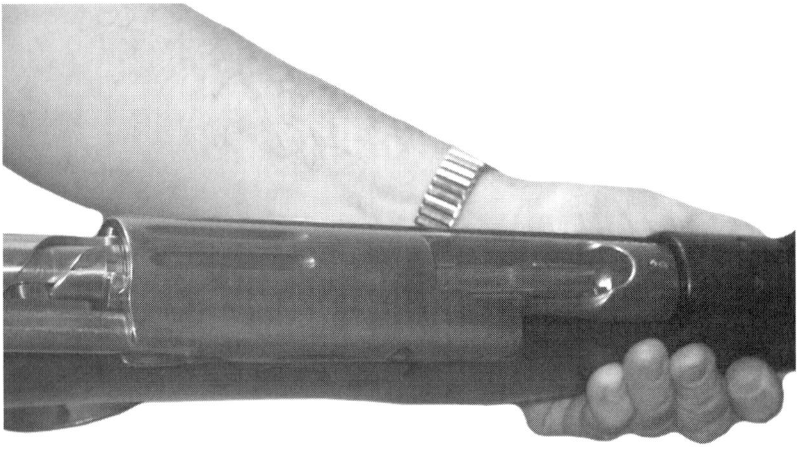

Fig. B-18. Open the bolt and look into the chamber and magazine to make certain that no cartridges are present.

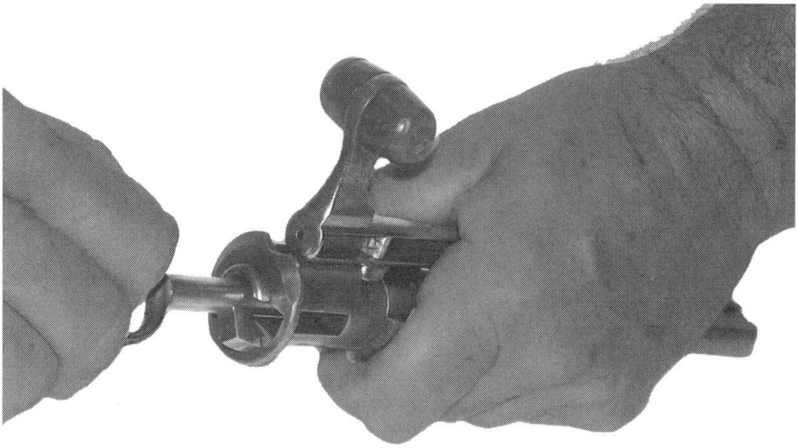

Fig. B-19. Pull the cocking piece back and turn clockwise slightly so that the cocking stud rests on the metal apron between the two rectangular notches. The bolt has been removed from the rifle for clarity.

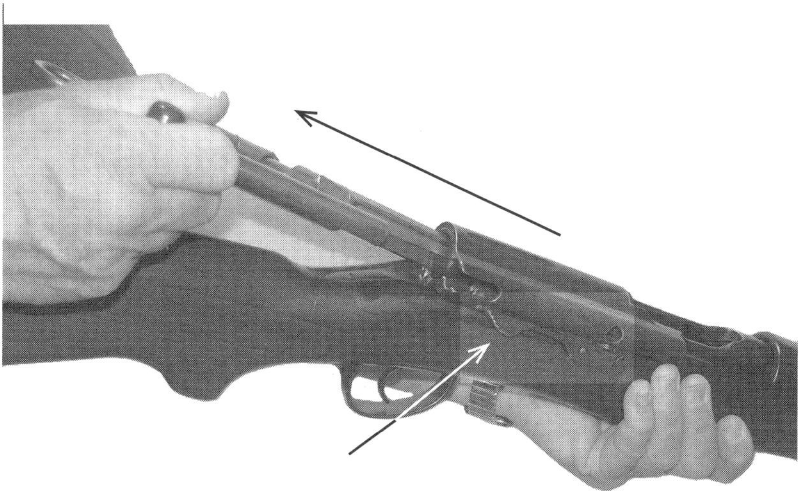

Fig. B-20. Depress the bolt release (arrow) and draw the bolt straight back out of the receiver.

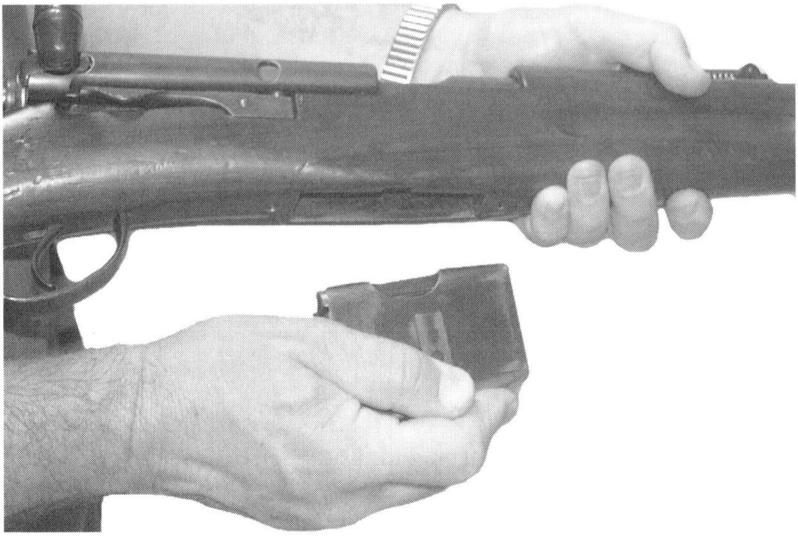

Fig. B-21. Remove the magazine from all rifles and carbines except for the Model 1889.

Fig. B-22. Loosen the clamping screw of the rear barrel band.

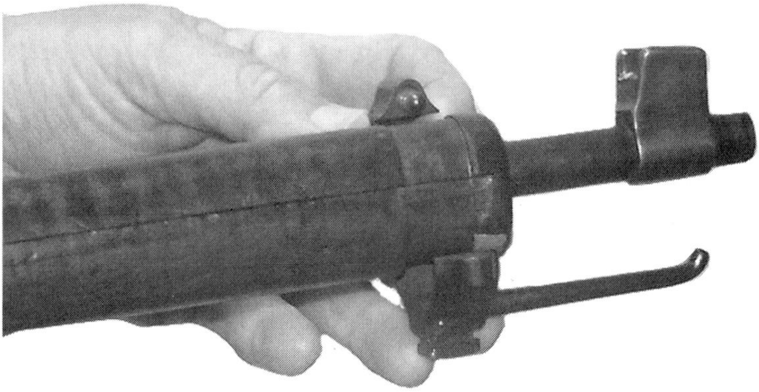

Fig. B-23. Loosen the clamping screw and remove the forward barrel band.

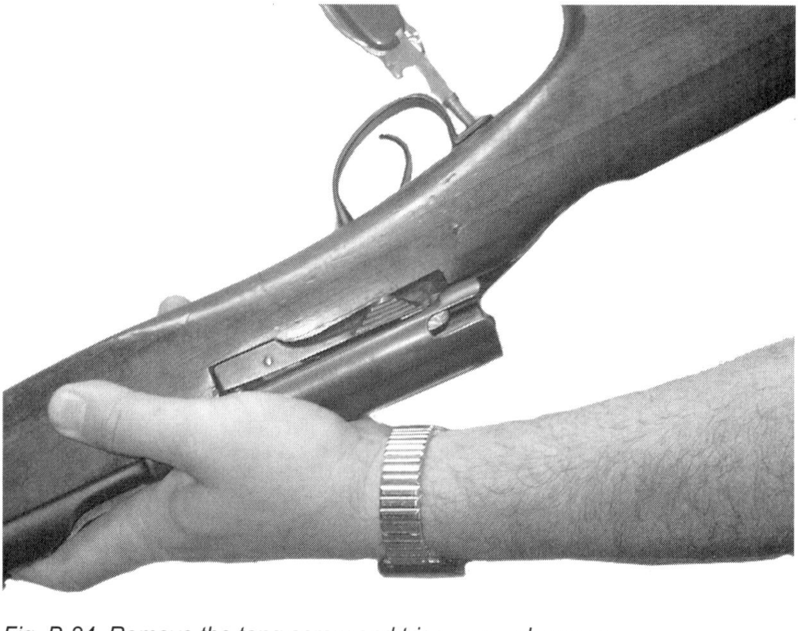

Fig. B-24. Remove the tang screw and trigger guard screws.

7. Remove the rear barrel band by depressing the band spring (if necessary) and sliding the band forward and off the rifle or carbine.

8. Remove the trigger guard screws (Figure B-24) and lay them aside in the order in which you removed them (four for the Models 1889, 1889/96 and 1889/00) (three for the Models 1905 and 1911) (two for the K31).

9. Remove the trigger guard plate, see Figure B-25.

First loosen it by gently rocking it from side to side. If it does not loosen immediately, treat the junction of wood and metal with lemon oil and wait for an hour or so, otherwise the wood will splinter along the edges.

10. The hand guard will lift straight off all rifle models and the K31. See Step 12 for removing the hand guard from the Model 1905 and Model 1911 carbines.

11. Remove the barreled receiver from the stock. To do so, lift the front of the barrel carefully to loosen it in the stock bed, then lift from the rear.

If barreled receiver is stuck, swab the areas of contact between metal and wood with lemon oil and wait an hour for it to penetrate. A heat lamp will hasten the process. After the lemon oil has loosened the dirt and natural varnish along the junction of wood and metal, hold the stock upside down and gently tap the barrel muzzle against a padded surface to loosen it and remove, see Figure B-26. Repeat the lemon oil treatment as necessary to avoid splintering.

12. The rear of the Model 1905 and Model 1911 carbine hand guards surrounds the rear sight. To remove, turn the barreled action upside down. Press on the spring clamp at the rear of the hand guard and push off the barrel. Turn the barreled action right side up. Lift the rear sight leaf. Rotate the hand guard so that it is perpendicular to the barrel and

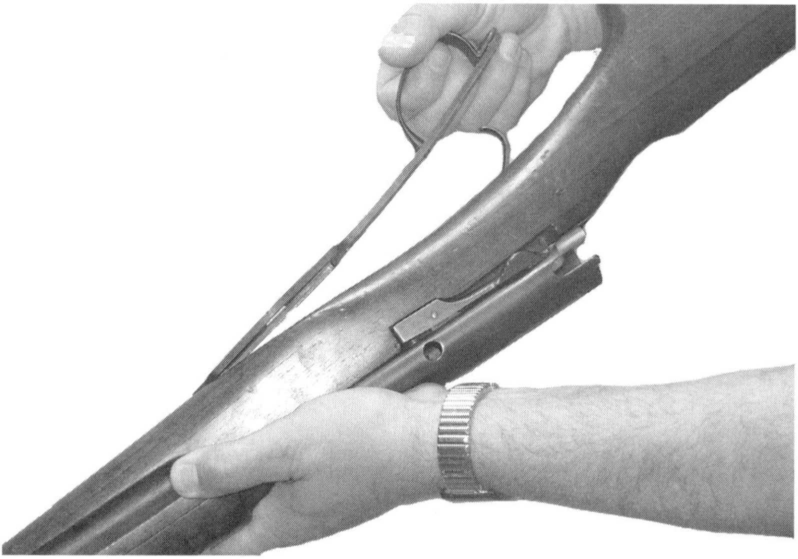

Fig. B-25. Gently rock the trigger guard from side to side to loosen, then lift.

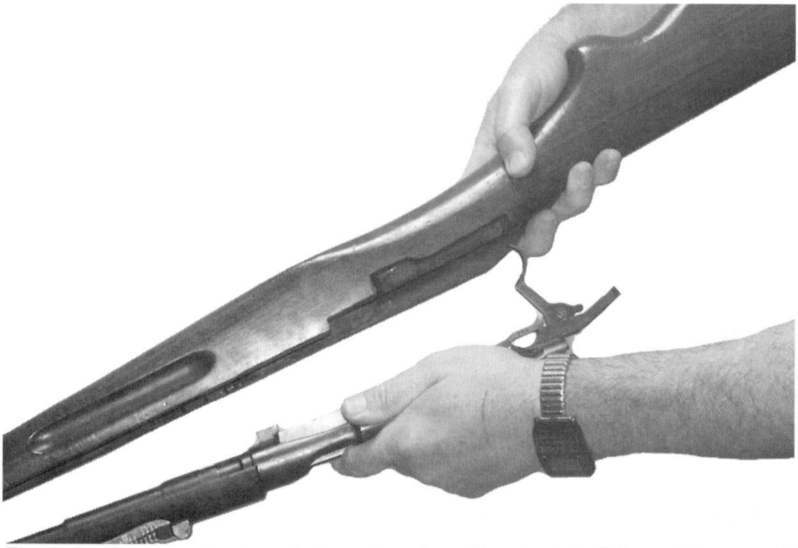

Fig. B-26. Remove the barreled receiver from the stock by lifting at the rear. If stuck, turn upside down, and tap the muzzle lightly on a padded surface.

lift off over the rear sight leaf. Reassemble in the same manner.

If you simply lift up on the rear hand guard to remove it from the barrel, it will probably split.

13. Remove the butt plate (Figure B-27). Use a thin bladed screwdriver, preferably European-style or the screwdriver blade of a Swiss Army knife, and unscrew the two butt plate screws. Rock the butt plate from side to side and pull off carefully. If the butt plate does not loosen, treat the junction of wood and metal with lemon oil. Also squirt a bit into the two screw holes.

14. The trigger assembly can be removed by driving out the trigger and sear pins but this is not recommended. If these parts need repair or adjusting, it should be done by a competent, trained gunsmith.

Reassembly is the reverse of the above instructions.

Bolt Disassembly and Reassembly

15. With the bolt removed from the receiver as in Step #2 and the cocking ring retracted so that the cocking lug is resting against the rear of the bolt cap, between the two slots, lay the bolt across your left palm with your thumb holding the locking lugs; push the bolt handle forward and lift up and off, see Figure B-28.

16. Unscrew the bolt cap from the bolt body by turning it counterclockwise until the parts separate, see Figures B-29 and B-30.

16a. To separate the parts of the K31 bolt, hold in the palm of your left hand with your index finger near the extractor.

16b. Turn the bolt plug a quarter turn clockwise and pull back. The bolt cap will separate from the bolt body.

16c. Tip the rear of the bolt body up and the bolt cylinder will slide out.

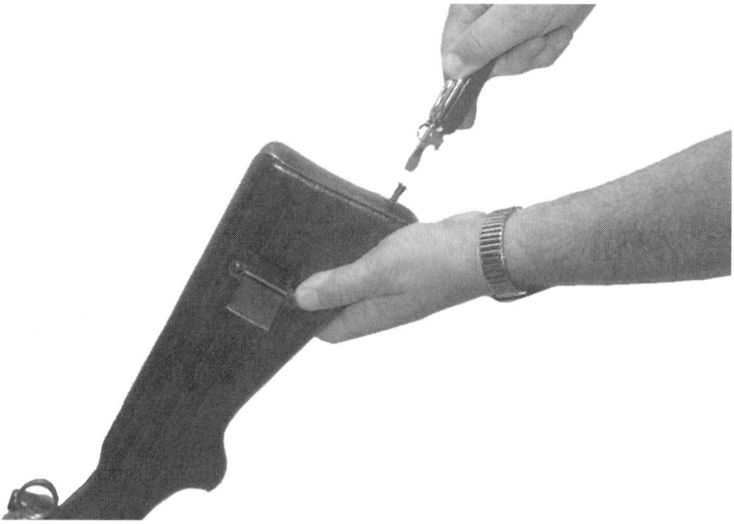

Fig. B-27. Remove the butt plate screws and tap the butt plate to loosen and remove.

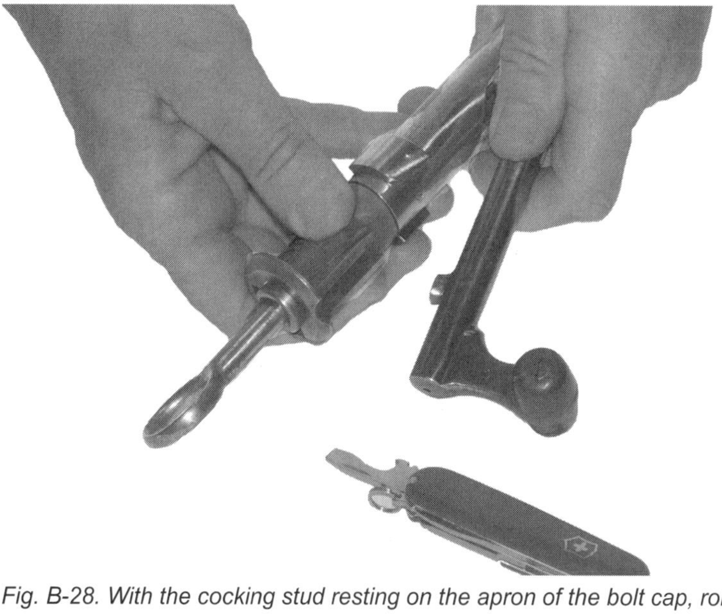

Fig. B-28. With the cocking stud resting on the apron of the bolt cap, roll the bolt body and cylinder until the cocking handle stud slides free.

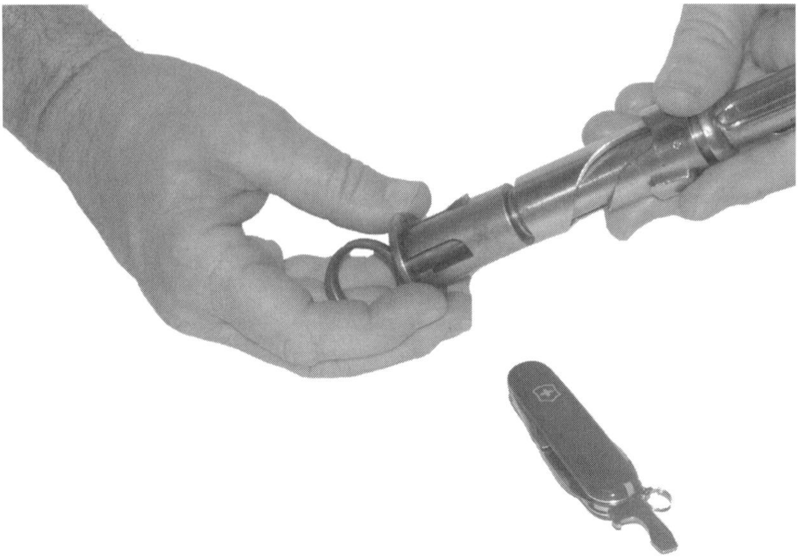

Fig. B-29. Unscrew the bolt cap from the bolt body.

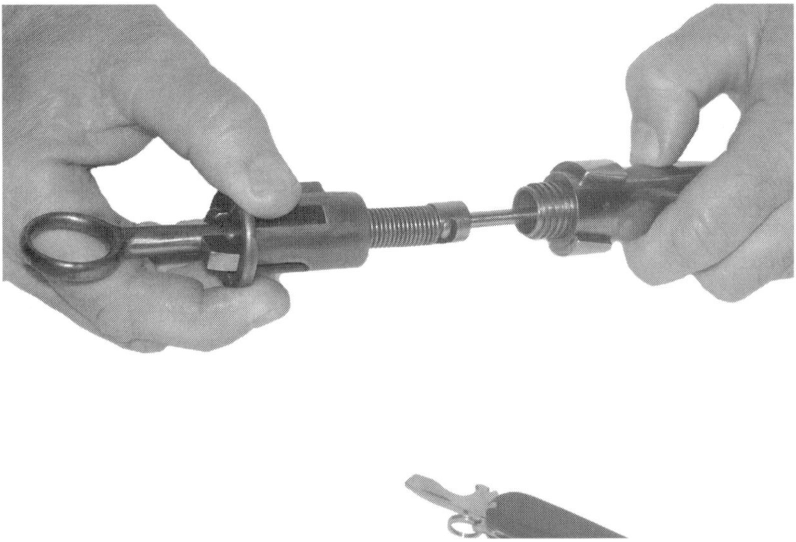

Fig. B-30. Remove the bolt cap with the striker assembly.

17. Hold the bolt cap with cocking ring in your left hand so that the cocking lug is up.

18. Pull back slightly and turn the cocking ring counterclockwise and ease the cocking lug into the long slot.

19. Holding the bolt cap in your left hand, cocking ring down, use your thumb and index finger to pull down on the bolt spring.

20. Withdraw the striker assembly from the bolt cap, refer to Figure B-30.

21. Pull the firing pin off the striker, noting how the attaching slot is oriented, see Figure B-31.

22. Separate the main spring from the striker.

23. Slide the bolt cylinder off the bolt body, see Figure B-32.

NOTE: The extractor should only be removed if replacement is necessary. It can be cleaned by squirting penetrating oil between the extractor and the bolt cylinder at the end and middle of the extractor.

24. If the extractor must be removed, place your thumbs at the right front and left rear and push so that the extractor rotates clockwise, see Figure B-33.

25. If it won't move, place the left front edge of the extractor against a block of wood or other solid surface and lightly tap the front of the bolt cylinder until the extractor snaps loose, see Figure B-34.

25a. The K31 extractor can be removed from the bolt body by pushing it out of its slot from the rear. Use a cartridge case rim or screwdriver tip, see Figure B-35.

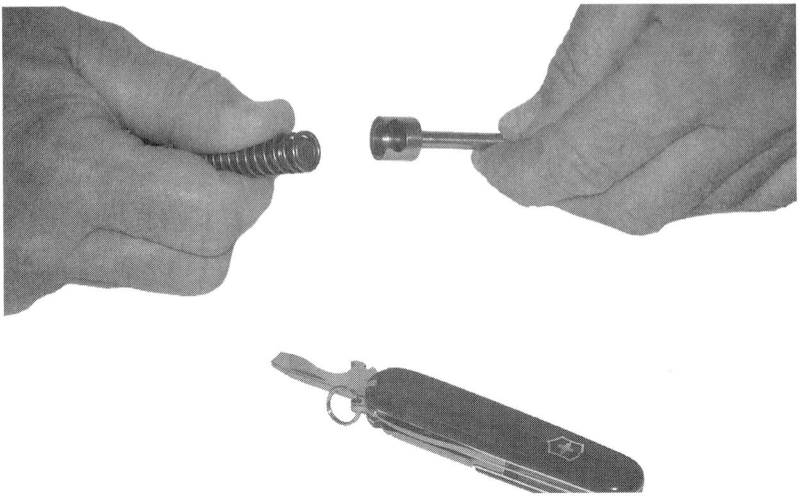

Fig. B-31. Remove the firing pin from the striker by compressing the bolt spring.

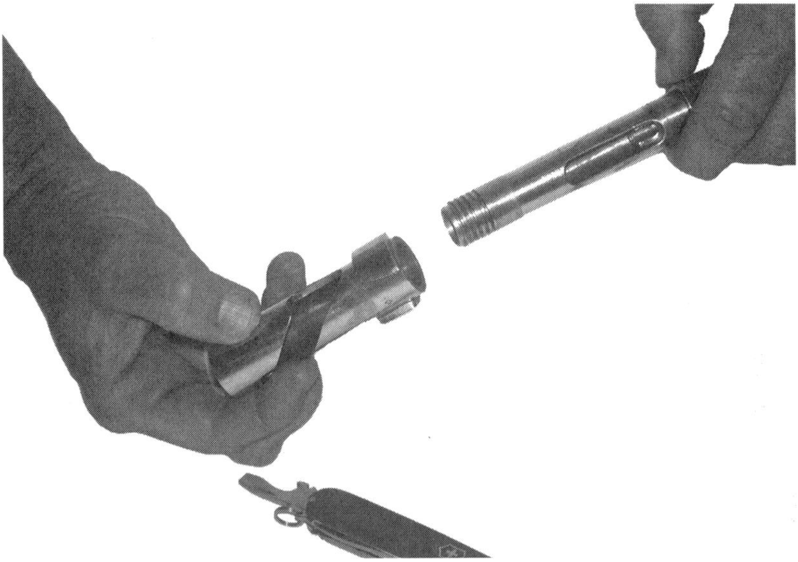

Fig. B-32. Tip up the bolt body and slide off the bolt cylinder.

25b. Be careful as the extractor is small and breaks easily.

26. To reinstall the extractor in the Model 1889 and Model 1911 bolt cylinders, turn the extractor, at right angles to the bolt cylinder, insert the pivot into its slot and rotate clockwise until it snaps into place.

26a. To reinstall the K31 extractor, carefully slide it back into its groove.

26b. Make certain that it is fully seated in its slot or the bolt will jam when you try and replace it in the bolt channel.

27. Removing the bolt face is not recommended. If it needs repair or replacement, it should be done by a competent, professional gunsmith with the proper tools.

28. Reassemble the bolt by reversing steps 15–26.

NOTE: When reinserting the bolt handle, line up the front of the curving bolt track on the bolt body with the slot on the bolt cap. Start positioning the bolt handle rear lug into the slot on the bolt cap slightly and let the front lug drop into the curving track on the bolt body. Rotate the bolt body counterclockwise to hold the bolt handle in place.

29. Reinsert the bolt into the bolt track in the receiver. The M1889 magazine cutoff switch should be up, lowering the magazine from the battery position. Push the bolt to forward to close and lock and set the safety by pulling back on the cocking ring and turning it clockwise into the safety slot. It is good practice to always set the safety to the "on" position.

30. Rotate the cocking ring 1/8th turn clockwise so that it slides into the safety slot. It is good practice to place the bolt in "SAFE" even if you intend to shoot the rifle or carbine immediately.

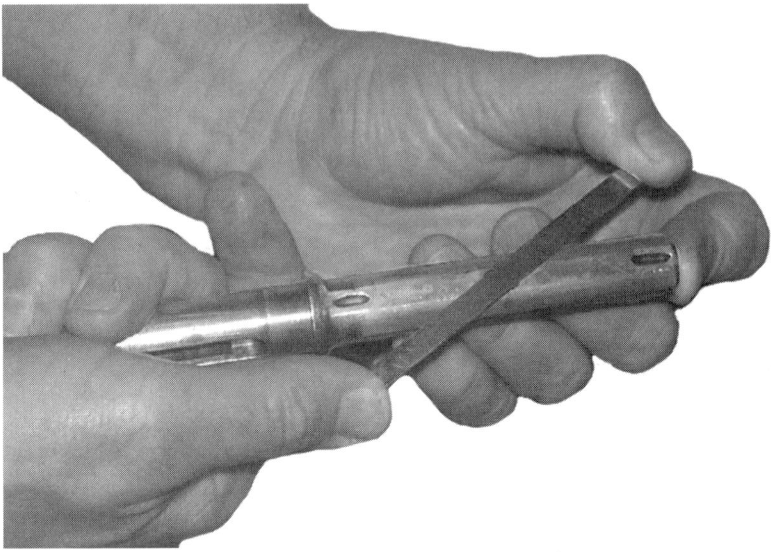

Fig. B-33. Place your thumbs on the top and bottom of the extractor and turn clockwise to remove.

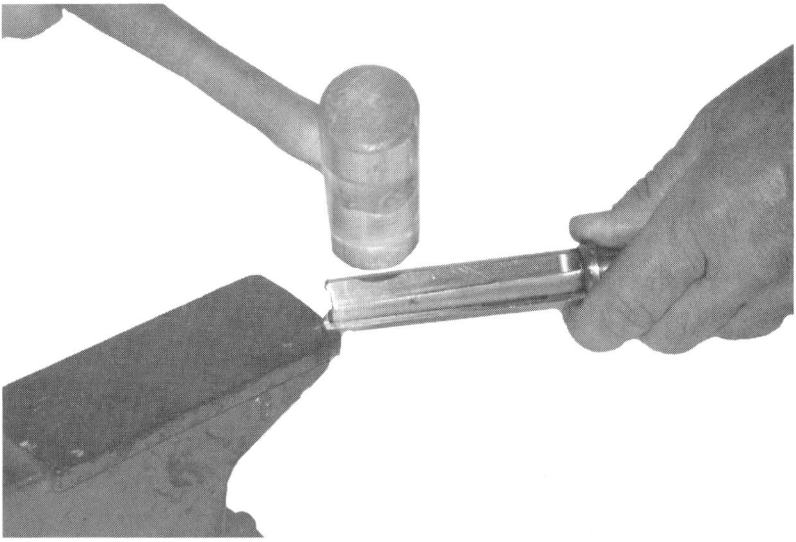

Fig. B-34. Or, place the top left side of the extractor on a hard surface and tap the bolt body with a non-marring hammer until the extractor snaps loose.

31. Replace the empty magazine in all models except the Model 1889 rifle.

Removing and Replacing the Model 1889 Rifle Magazine

To remove the magazine from the Model 1889 rifle, the trigger guard plate must be removed from the stock.

32. Remove the magazine cutoff switch pivot screw.

33. Press the magazine cutoff switch down (arrow A), then press back against the magazine body and pivot counterclockwise below the trigger guard plate to draw its lug (arrow B) out of the notch in the bearing plate (C) on the magazine's right side, see Figure B-36. The plate with the bearing notch was made just small enough that the lug on the magazine cutoff switch will pass by its left side when the switch is pivoted as far down (counterclockwise) as possible.

34. Carefully push the magazine up through the trigger guard.

35. The Model 1889 and all other Schmidt magazines can be disassembled by pushing down on the magazine follower, tipping it to the left and wriggling it up through the magazine's feed lips. The Z-spring can be removed the same way.

36. To reassemble, reverse the procedure in steps 32 through 35, taking care to insert the magazine cutoff switch first. Insert the pivot screw after the magazine is in place. To reinsert the trigger guard and magazine into the stock, the magazine cutoff switch must be down.

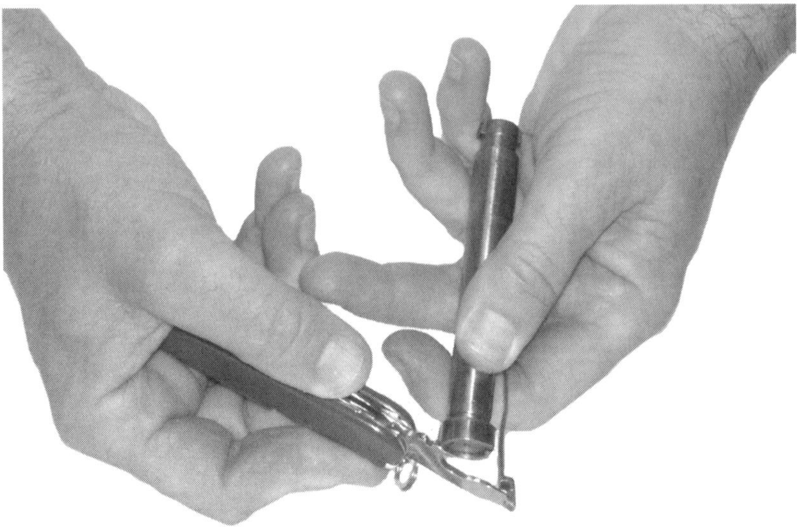

Fig. B-35. Remove the K31 extractor by carefully pushing its hook up so that its securing stud clears its detent, and slide forward. Do not bend it by pushing up too far.

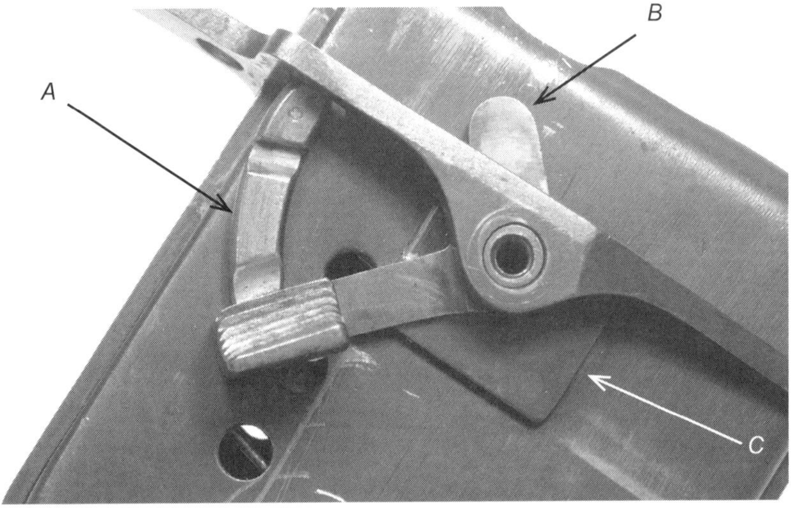

Fig. B-36. Press the magazine cutoff switch (A) down, then back against the magazine and rotate counterclockwise to draw its lug (B) out of the notch in the bearing plate (C).

Appendix C

Shooting and Maintaining the Vetterli and Schmidt Rifles

General Considerations

Make certain that your rifle or carbine is in good physical shape and that you have disassembled the bolt and inspected and cleaned its parts. Check the firing pin tip and the locking lugs for cracks; use a strong magnifying glass. Reassemble the striker/firing pin without the bolt spring and make certain that it moves smoothly and that the firing pin protrudes just enough to make contact with the primer in the cartridge case.

Inspect the trigger/sear mechanism and the safety system to make certain that both function properly and that 1) the sear holds the cocking piece securely when the bolt is closed and 2) that the sear is not worn so that the cocking piece can slip. 3) Make certain that the Vetterli's magazine tube is clean and dry and that the internal parts function freely.

It is an excellent idea to have any collectible firearm inspected by a trained gunsmith. In fact, the older the firearm, the more important a thorough inspection becomes.

The Vetterli System

Each Vetterli rifle and carbine uses a magazine tube running through the forend or forearm. It cannot be loaded with a clip or other feeding device. The Vetterli cartridge is described as 10.4 x 41 mm, meaning that the nominal diameter of the bullet is 10.4 mm (.41 caliber) and the cartridge case is 41 mm long, see Figure C-1. Vetterli bullets are wrapped in a paper patch which is lubricated with a wax to reduce fouling. The propellant was black powder. In North America, the cartridge was manufactured by a variety of U.S. and Canadian ammunition firms

Fig. C-1. The 10.4 x 41 mm Rimfire Vetterli cartridge.

at various times and it was usually known as the .41 Swiss.

Cartridges of the proper caliber are inserted one at a time through the loading port in the right side of the receiver, see Figure C-2. The bolt must be closed so that the cartridge lifter is in the down position. The tubular magazine in a Vetterli rifle will hold either eleven (Model 1869/71) or twelve cartridges (Model 1878/81). Consult Chapter 1 for the magazine capacities of the various carbines.

It is a good idea not to load to capacity as old springs lose some of their elasticity.

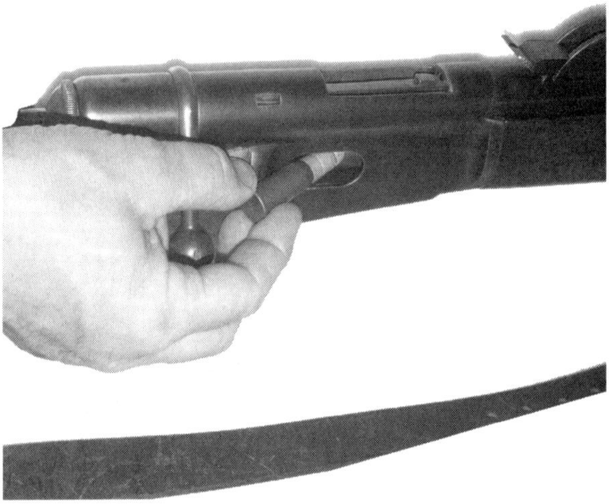

Fig. C-2. The Vetterli is loaded one cartridge at a time through a port in the receiver.

NOTE: Vetterli rifles and carbines do not have a manual or other safety system. A common misconception is that with the bolt handle lifted but the bolt not withdrawn, the locking lugs are rotated so that the sear is not in contact with the cocking lug and the rifle cannot therefore be fired. But closing the bolt only as much as ten or twelve degrees will cock the firing mechanism and a sudden jar or pulling the trigger inadvertently will allow the rifle to fire with the bolt improperly closed. The shooter or a bystander could be severely injured or killed if this happens.

After the bolt is closed, the striker will protrude from the bolt cap as a visual indication that the rifle is cocked. You should also leave the bolt fully open and withdrawn from the breech after loading until you are at the firing line, ready to shoot. Immediately after shooting, turn the rifle upside down and unload any unfired cartridges by carefully working them through the action by opening and closing the bolt. The cartridges will fall out of the elevator. Inspect the end of the magazine tube visible through the top of the breech with the cartridge lifter in the down position to make certain that the magazine is entirely empty and that a cartridge is not hung up in the tube because the cartridge lifter is not fully down, see Figure C-3.

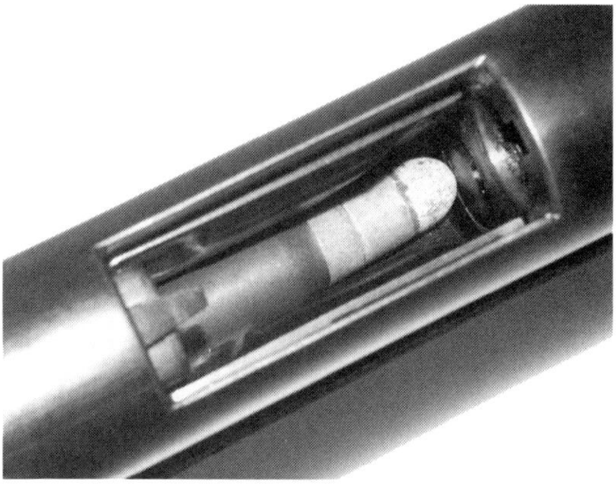

Fig. C-3. Always look down into the open top of a Vetterli receiver to make certain that a cartridge is not hung up in the cartridge lifter.

In all Vetterli rifles but the Model 1869, the rear sights are graduated in meters. The Model 1869 sights are graduated in "Schritt" or paces which are close enough to meters not to matter with the quality of most ammunition you might find today.

The Schmidt System

The Schmidt rifles all fired the 7.5 mm Swiss cartridge in a small number of variations, see Figure C-4. Refer to Chapter 2 for a discussion of the 7.5 mm ammunition and the variations to be used in specific models.

The Schmidt magazines can be loaded from the specialized paper and tin clips (although these are getting very hard to find) manufactured for the Models of 1889 and 1911 or from loose cartridges, see Figure C-5. Model 1889 magazines hold twelve cartridges which, unless you have clips, must be pressed down one at a time through the open breech. Model 1911 and K31 magazines can be removed to load six cartridges at a time, see Figure C-6, or can be loaded with the magazine in the rifle by placing the clip in the guides exposed when the bolt is open and pressing down on the back of the top cartridge with your thumb. Press down smoothly and make certain that the last cartridge rests below the magazine's feed lips.

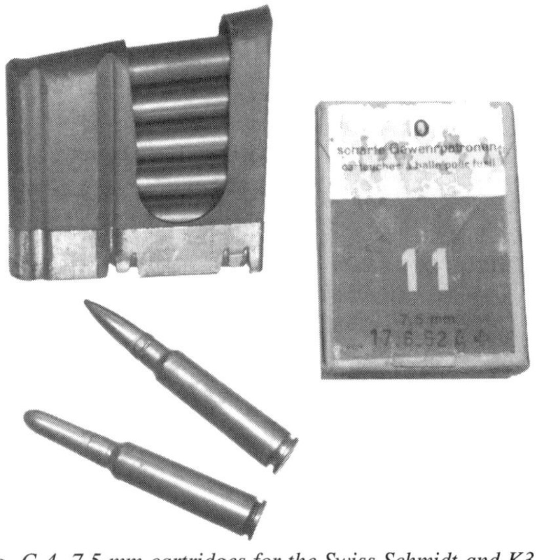

Fig. C-4. 7.5 mm cartridges for the Swiss Schmidt and K31 rifles: clockwise from top left, GP 1890 (7.5 x 53.5 mm) in clip, GP11 (7.5 x 55 mm) in ten-round box, GP11 cartridge and GP90/23 (7.5 x 54.5 mm).

The Model 1889 rifle will only feed cartridges from the magazine if the magazine cutoff is "off" (cutoff switch down). Only the Model 1889 rifle has a magazine cutoff switch.

Fig. C-5. *Two clips of six cartridges are used to load the Model 1889 rifle through the open breech.*

Load the magazine. Before inserting into the rifle, make certain that the bolt is closed and the safety is in the "on" position. Insert the magazine.

When you are on the firing line and ready to shoot, turn the safety to the "off" position and work the bolt to load the first cartridge. Unless you are planning to shoot immediately, reset the safety to "on."

To unload any remaining cartridges in the magazine, in the Model 1889, carefully work the bolt until the magazine is empty. The Model 1911 and K31 magazines may be removed and emptied by pushing out the cartridges one at a time.

It is a good idea when storing the rifle with the magazine inserted, to leave the bolt open. Anyone handling the rifle at a later time will be able to see immediately that the rifle is empty.

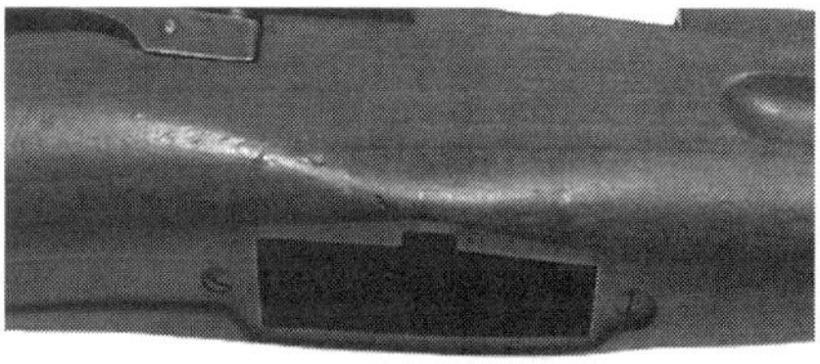

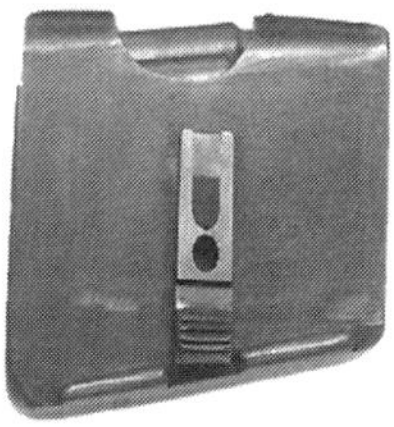

Fig. C-6. The Models 1889/96, 1889/00, 1905, 1911 and K31 may be loaded through the open breech using the clip or by removing the magazine and loading with loose cartridges.

Maintenance

Regular cleaning and inspection is the key to a reliable and safe firearm. If you are using black powder cartridges or corrosive ammunition, proper cleaning as soon after shooting, and certainly within six hours, is extremely important to prevent rust and corrosion from destroying the accuracy.

To clean black powder fouling or the salts from corrosive primers, dismount the action from the stock. Remove the bolt. Boil one quart of water with a teaspoon of dishwashing detergent and a second quart of plain water. Stand the barreled action in a pail or other container. Place the disassembled bolt parts in the bottom of the pail. Use a funnel in the muzzle and padded pliers or wrench to hold the barrel upright and pour most of the quart of the boiling soapy solution through the bore. Wind a soft cloth around a .22-caliber cleaning rod to make a plug and insert through the bore into the chamber. Fill the bore with some of the remaining hot soapy water and pump up and down the length of the bore. Repeat several times. Turn the barreled action over and repeat with special attention to the chamber.

Pour most of the plain hot boiling water through the muzzle slowly to rinse and set aside to dry.

Clean the bolt parts by scrubbing with a soft brush or cloth. Rinse with the remaining hot plain water and set aside to dry.

Oil all parts and the bore and chamber with a water-displacing oil, dry and coat with a thin film of gun oil and reassemble.

Reinspect in two or three days.

Firearms using modern smokeless powders need much less attention. To clean, simply follow the manufacturer's instruction on the cleaning solvent being used. Pay particular attention to bores, chambers and bolt faces.

Since smokeless powders do not cause rust or corrosion, there is no need to scrub bores with stiff brushes which will only wear the land and the delicate bore at the muzzle, destroying accuracy. Use a brass or wooden cleaning rod (avoid steel rods as they wear the muzzle, or use a chamber and bore guide) and soak a patch two inches square or less with the cleaning solvent. Remove the bolt. Wrap the patch around the cleaning rod tip and run it through the bore *from the chamber end.* Let the patch fall off when it exits the bore. Withdraw the rod and mount another patch. Do not run the same patch back and forth. Repeat five or more times until the patch comes out clean. Use a chamber brush to clean fouling and unburnt powder from the breech and wipe with a solvent-saturated patch. Run a dry patch through the bore, then one more patch treated with gun oil. Do not cap or plug the muzzle or moisture will condense inside the barrel and cause rust. If you use a muzzle cap to prevent dust from settling in the bore, make certain that it is loose fitting and allows air to circulate freely.

Lightly oil bolt and trigger parts. A firearm's worst enemy is over-oiling. Moving parts should receive a very light coat of oil applied from an oil-soaked cloth. If you need to rub down the exterior metal parts to remove fingerprints and dirt, use a silicone-treated gun cloth available at any firearms dealer. Silicone helps prevent rusting.

Store firearms in a clean, dry place away from dust and grit. It is best to store firearms in a horizontal position whenever possible. Place a porous cloth over open actions or muzzles, if upright, to protect from dust. Inspect periodically for rust or corrosion, especially in the winter when temperatures in homes cycle through significant variations from day to night.

Note: It is always a good idea to wear eye protection when disassembling and/or cleaning a firearm. You only have one pair of eyes.

NOTE: Swiss firearms were designed to work under abnormal temperature extremes. The K31, particularly, was designed to be fired in very cold winter temperatures without lubrication, but only under combat conditions. Operating any firearm without proper lubrication should only be done under extremely frigid conditions, and then only for as long as necessary.

In civilian sporting use, this translates to never.

Appendix D
Troubleshooting
the
Vetterli and Schmidt Rifles

Vetterli System

Most Vetterli problems fall into three categories: first, the rifle will not fire. This is almost always caused by the fact that virtually all ammunition now available for the rimfire Vetterli system is at least 60 years old and most is more than 100 years old. The priming compound used in rimfire cartridges is often unstable over long periods and will break down.

Secondly, a broken firing pin. The twin prong rimfire firing pin is not overly delicate but has been known to break. If you cannot secure a replacement, you can easily make one with a hacksaw and file. First check under the butt plate for the spare firing pin placed there by the factory.

If it is gone, use the old firing pin as a pattern. Cut a firing pin from mild steel plate of the same or slightly greater thickness. File to shape. Make certain that the firing pin slides as easily into the slot in the bolt as the old pin.

The new firing pin must be stress relieved and then heat-treated. The former can be accomplished by heating the new firing pin to dull red for thirty seconds or so and then allowed to cool. To heat-treat, place the firing pin on a piece of steel plate and heat the plate until the firing pin glows cherry red. Hold at that temperature for one minute, maintaining the same color (a process called soaking), then slide the firing pin off the steel plate into a container holding at least one quart of water. When cool, polish with steel wool.

The third category of problems has to do with worn locking lugs. Although rare, the only fix is to replace the entire bolt handle and then have a competent gunsmith check the headspace. Head space settings in most rimfire firearms depend on the thickness of the cartridge rim, so be sure and supply the gunsmith with a sample cartridge.

Another possibility is dirt and grease in the bolt blocking the action of the firing pin. To check, disassemble the bolt following the instructions above and clean the bolt, paying particular attention to the striker tunnel in the bolt body and the firing pin.

If cartridges refuse to feed from the magazine, the magazine tube is either dirty or full of dried grease. To clean, remove the forend or forearm and unscrew the knurled ring that holds the follower in place.

Schmidt System

Problems with the Schmidt rifle are usually because 1) the shooter has not closed the bolt completely, 2) the firing pin tip is broken or 3) the owner has allowed dirt and grease to accumulate.

If the rifle misfires, immediately point the rifle in a safe direction down range. Set the safety by turning the cocking piece clockwise to the 9 o'clock position. Wait one minute to make certain that the cartridge is not a "hang fire" (delayed ignition).

Release the safety and open the bolt. Remove the cartridge and inspect the primer for a solid strike by the firing pin. If the primer is barely dented, the firing pin is probably stuck in position by dirt or dried grease or is broken.

Disassemble the bolt following the instructions in Appendix B. Clean the bolt parts, paying particular attention to the striker tunnel. Make certain that the firing pin tip is not broken and protrudes just enough to contact the primer in the cartridge case.

If the bolt will not open, reset the safety. Remove the magazine (magazine removal in the Model 1889 requires removing the trigger guard assembly).

Release the safety once more and pull back on the bolt handle. If the bolt still does not move, strike the bolt handle a sharp backward blow with the palm of your hand.

If the bolt still does not open, the services of a gunsmith may be required.

Careless loading of cartridges may cause the rifle to misfire. When inserting the cartridges into the magazine, take care to see that they are pushed down past the feed lips, particularly the sixth and last cartridge.

In the rare instance of an empty cartridge case becoming stuck in the chamber, try opening and closing the bolt several times. If that does not work, remove the bolt. Rest the buttstock on a solid surface with the rifle vertical. Using a cleaning rod or similar stiff metal rod small enough in diameter to pass through the bore, insert through the muzzle until it rests on the bottom of the stuck case. Lightly tap the end of the rod until the case is dislodged. Have patience and do not strike a hard blow. This will only deform the case and increase friction, making it even harder to remove the case.

If this procedure does not work, the services of a gunsmith may be required.

Spare Parts

The collector and shooter should keep in mind that spare parts for the Vetterli are almost nonexistent and those for the Schmidt models, including the K31, are scarce outside Switzerland. When the opportunity offers, purchase spare parts, particularly firing pins and springs. The natural ruggedness and the excellent Swiss craftsmanship do minimize the need for spare parts.

Finally, use only ammunition from well-known ammunition manufacturers. If you use surplus military ammunition, be sure you match the 10.4 mm Vetterli or the 7.5 mm Swiss cartridges to the model rifle or carbine you are shooting. Complete descriptions, including dimensions, are found in Chapters 1 and 2.

Appendix E
Glossary

Aperture	Opening in the rear sight.
Blueing	A chemical or heat process by which the top layer of steel or iron is oxidized to a blue-black color. The objective is to darken the metal to prevent glare rather than retard rusting.
Bore	Hollow interior of a gun barrel.
Braze	Process by which two metal parts are joined together by allowing a mixture of melted brass and silver to flow between two joints when heated.
Breech	The rear end of a gun barrel or forward end of the receiver that holds the cartridge to be fired.
Butt	The rear end of a gun stock that is held against the shoulder.
Butt Plate	Metal cap that covers the end of the buttstock to prevent splintering.
Cannelure	Groove around the base of a rimless cartridge case.
Canton	Similar to states. Cantons are political subdivisions of the Swiss federation.
Carburize	The process by which carbon is driven into steel by heat to harden it.

Cartouche	Letters, numbers or a symbol inside a frame and cut into wood or metal with a die. Used in gun manufacturing as a mark of acceptance.
Case-Harden	A process in which iron or low-carbon steel is heated to red hot in the presence of a source of carbon, then cooled quickly without exposure to air to form a thin surface layer or hard metal over a softer, less brittle core. If the metal is cooled in water, it produces mottled blues, reds and yellows. If cooled in oil, a black color is imparted to the metal.
Cleaning Rod	One-piece or jointed wood or metal rod used to clean the barrel's bore.
Clip	A device for holding a number of cartridges in line so that they can be easily and quickly loaded into a firearm.
Comb	The top ridge of the buttstock.
Crown	The rounded or recessed end of the muzzle that protects the rifling.
Drift	Flight to right or left of the line of sight caused by the spin imparted to the bullet by the rifling.
Ferrule	Metal ring or cap used to strengthen a thin part.
Forearm	Part of the carbine stock forward of the receiver, under the barrel.
Forend	Part of the rifle stock forward of the receiver, under the barrel.

Helvetic Republic	Informal name for Switzerland, after the Latin Helvetia, the name for the Roman Empire's province containing what is today, Switzerland.
In The White	Steel or iron parts left in their natural (white) color without the application of any other finish.
Lands	Raised spirals inside the barrel which impart spin to the bullet.
Left Side	Side opposite the bolt handle.
Lightening Cuts	Wood or metal removed from a part to make it lighter.
Lozenge	A mark cut into wood or metal having two parallel lines connected at either end with an arc. Usually containing letters, numbers or symbols and used as an indication of manufacture, ownership or acceptance.
Lug	Ear-like projection which acts as a support, or to which a support can be fastened.
Milled	To shape or dress metal using a rotary cutter.
Muzzle	The front end of the barrel.
Needle Fire	Refers to the long, needlelike firing pin of the Dreyse rifle, which penetrated the length of a paper cartridge before striking the priming compound at the front where it rested against the base of the bullet.
Proof Mark	Signature initial(s), number or other mark stamped

	on wood or metal, signifying that the part or completed arm has been inspected and tested and meets established specifications.
Refurbish	To rebuild or otherwise restore a firearm to full service.
Right Side	Same side as the bolt handle.
Sans Serif	A type face in which letters or numbers are without short strokes depending at an angle from the main stroke of a number or letter. This sentence is an example of a sans serif type face.
Serif	A type face in which the letters or numbers do have short strokes depending at an angle from the main stroke of number or letter. This sentence is an example of a type face with serifs.
Stud	A projection from a surface to which another part is attached.
Temper	To change the physical properties of a metal by heating.
Zero	To adjust the sights of a firearm so that the bullet strikes the center of a target at a given distance.

Appendix F
Bibliography

____, *Arms of the World, 1911*, ALFA Catalog, Follett Publishing Company, Chicago, IL, reprint of 1911 catalog.

_____, *Bannerman Catalog, The*, Francis Bannerman Sons, Inc., NY, 1904.

_____, *Bannerman Catalog, The*, 100th Anniversary, Francis Bannerman Sons, Blue Point, NY, 1966.

Barnes, Frank. *Cartridges of the World,* DBI Books, Northbrook, IL 60062, 8th Edition, 1997.

Datig, Fred A. *Cartridges for Collectors,* Vols. 1, 2, 3, 4, Borden Publishing Company, Alhambra, CA 91801, 1956 and later.

Ezell, Edward C. *Small Arms of the World*, Arms and Armor Press, Stackpole Books, Harrisburg, PA. 12th revised edition, 1983.

Frasca, Albert J., and Robert H. Hill. *1909 Catalog, Springfield Armory Museum Arms & Accouterments*, Springfield Publishing Company, Carson City, NV 89702, 1995.

Hogg, Ian V. and John Weeks. *Military Small Arms of the 20th Century,* DBI Books, Northfield, IL 60093, 4th Edition, 1981.

Janzen, Jerry L. *Bayonets from Janzen's Notebook*, PO Box 2863, Tulsa, OK 74107, 1984.

Johnson, George B., and Hans Bert Lockhaven. International Armament, Volume I, International Small Arms Publisher, Cologne, Germany, 1965.

Otteson, Stuart. *The Bolt Action Rifle*, Winchester Press, New York, NY, 1976.

Rubin Rifle; Report of the Secretary of War: Message and Documents, Two Houses of Congress, 1st Session, 51st Congress. Volume III, Washington, 1889.

Sallaz, Kurt, and Michael am Rhyn. *Handfeuerwaffen Gradzug-Systeme*, Bewaffnung und Ausrüstung der Schweizer Armee Seit 1817, Verlag Stocker-Schmid, Dietikon-Zürich, Switzerland, 1978

Scarlata, Paul S. *Collecting Classic Bolt Action Military Rifles*, Andrew Mowbray Publishers, PO Box 460, Lincoln, RI 02865, 2001.

Smith, W.H.B., and Joseph E. Smith, *The Book of Rifles*, The Stackpole Company, Harrisburg, PA, 3rd Edition, 1963.

Tanner, Hans. *Guns of the World*, "Swiss Military Arms," Petersen Publishing, Los Angeles, CA, 1977.

Wallace, Captain W. B. *Textbook of Small Arms*, HMS Stationery Office, London, 1904.

Webster, Donald B. *Military Bolt Action Rifles, 1841-1918*, Museum Restoration Service, Bloomfield, Ontario K0K 1G0, 1993.

Wirnsberger, Gerhard, and R.A. Steindler (trans.). *The Standard Directory of Proof Marks*, Published by Jolex, Inc., in cooperation with The John Olson Company, Paramus, NJ 07652, 1975.

About the Author

Joe Poyer is the author of more than 400 magazine articles on firearms, the modern military, military history and personal security. He has written and published twelve novels with worldwide sales exceeding five million copies and authored or coauthored nine nonfiction books on the modern military from other publishers.

He is the editorial director and publisher of North Cape Publications®, Inc., which publishes the "For Collectors Only®" and "A Shooter's and Collector's Guide" series of books for firearms collectors and shooters. In these series, he has written or coauthored: *The .45-70 Springfield*; *U.S. Winchester Trench and Riot Guns, and Other U.S. Combat Shotguns*; *The M1 Garand, 1936 to 1957*; *The SKS Carbine*; *The M14-Type Rifle*; *The SAFN-49 Battle Rifle*; *The Swedish Mauser Rifles*; *The M16/AR15 Rifle*; *The Model 1903 Springfield Rifle and Its Variations*; *The American Krag Rifle and Carbine*; *Swiss Magazine Loading Rifles, 1869 to 1958*; *The AK-47 and AK-74 Kalashnikov Rifles and Their Variations;* and *The Model 1911 and Model 1911A1 Military and Commercial Pistols*.

Mr. Poyer has served as editor of the following magazines: *Safe & Secure Living*; *International Military Review*; *International Naval Review* and as field editor for *International Combat Arms*. He is currently at work on a new book in the "For Collectors Only" series, *The American Sniper Rifle*.

Mr. Poyer was the on-camera Military Affairs Analyst and Reporter for a major television station in Los Angeles, California. He also imported the very fine L1A1A inch pattern FAL rifles from Australia in the late 1980s.

Publications Available from

North Cape Publications®, Inc.

The books in the "For Collectors Only®" and "A Shooter's and Collector's Guide" series are designed to provide the firearms collector with an accurate record of the markings, dimensions and finish found on an original firearm as it was shipped from the factory.

For Collectors Only® Series

The Model 1911 and 1911A1 Military and Commercial Pistols, by Joe Poyer ($35.95). The first new book on the Model 1911/1911A1 in over fifteen years. Provides a complete part-by-part by serial number range for both military and commercial models as well as several foreign models. Serial number charts are extensive and cover each variation individually. Additional charts compare military and serial numbers by year of manufacture and provide an annotated list of military serial numbers. All markings—factory inspection, military acceptance, and factory address and patent markings—are discussed and illustrated.

M1911/1911A1/Government Models, Photographic Images: Pistols, Model 1911, 1911A1, and Government Models, *The Karl Karash Collection*, 2004.
Karl Karash has produced a portfolio on Compact Disk of over 5,000 color photographs of Model 1911/1911A1 pistols, both military and commercial, which is available from North Cape Publications, Inc. This collection is an invaluable tool for researching the M1911/M1911A1 Pistol. A total of 205 pistols are shown full size in right and left views, and in close-ups of all significant parts and markings. Available from North Cape Publications, Inc., PO Box 1027, Tustin, CA 92781. $19.95 plus $2.95 postage.

The .58- and .50-Caliber Rifles and Carbines of the Springfield Armory, 1865–1872, by Richard A. Hosmer ($19.95). This book describes the .58- and .50-caliber rifles and carbines that were developed at the Springfield Armory between 1865 and 1872 and which led ultimately to the selection of the famed ".45-70 trapdoor."

Serbian and Yugoslav Mauser Rifles, by Branko Bogdanovic ($19.95). Thousands of Yugoslav Mauser rifles have been imported into North America in the last two decades and Mr. Bogdanovic's book will help the collector and shooter determine which model he or she has, its antecedents, andwhether or not it is original as manufactured. Every Mauser that found its way into military service in Serbia/Yugoslavia from the 1870s to the present is listed and described.

Swiss Magazine Loading Rifles, 1869 to 1958 (2nd edition, revised), by Joe Poyer ($22.95). A complete part-by-part description for the Vetterli, Schmidt-Rubin, and K31 rifles in all their variations by serial number range, plus a history of their development and use, their cleaning, maintenance, and how to shoot them safely and accurately.

The American Krag Rifle and Carbine (2nd edition, revised), by Joe Poyer, edited by Craig Riesch ($22.95). A part-by-part description of the first magazine repeating arm adopted for general service in American military history.

The Model 1903 Springfield Rifle and Its Variations (3rd edition, revised and expanded), by Joe Poyer ($24.95). Includes every model of the Model 1903 from the ramrod bayonet to the Model 1903A4 Sniper rifle. Every part is described by serial number range, markings and finish.

The .45-70 Springfield (4th edition, revised and expanded), by Joe Poyer and Craig Riesch ($19.95), covers the entire range of .45-caliber "trapdoor" Springfield arms, including bayonets, tools and accoutrements.

U.S. Winchester Trench and Riot Guns and Other U.S. Combat Shotguns (2nd edition, revised), by Joe Poyer ($16.95). Describes the elusive and little-known "Trench Shotgun" and all other combat shotguns used by U.S. military forces.

U.S. M1 Carbines, Wartime Production (6th edition, revised and expanded), by Craig Riesch ($22.95), describes the four models of M1 Carbines from all ten manufacturers. Complete with codes for every part by serial number range.

The M1 Garand, 1936 to 1957 (5th edition, revised and expanded), by Joe Poyer and Craig Riesch ($22.95). This book covers such important identification factors as manufacturer's markings, proof marks, final acceptance cartouches stampings and heat treatment lot numbers plus detailed breakdowns of every part in minute detail.

Winchester Lever Action Repeating Firearms, by Arthur Pirkle
Volume 1, **The Models of 1866, 1873 & 1876** ($22.95)
Volume 2, **The Models of 1886 and 1892** ($22.95)
Volume 3, **The Models of 1894 and 1895** ($22.95)
These famous lever action repeaters are completely analyzed part-by-part by serial number range in this first new series on these fine weapons in twenty years.

The SKS Carbine (4th revised and expanded edition), by Steve Kehaya and Joe Poyer ($19.95). The SKS Carbine "is profusely illustrated, articulately researched and covers all aspects of its development as well as . . . other combat guns used by the USSR and other Communist bloc nations." Glen Voorhees, Jr., *Gun Week*.

British Enfield Rifles, by Charles R. Stratton (Volume 1 and, $19.95)
Volume 1, **SMLE (No. 1) Mk I and Mk III** (3rd edition)
Volume 2, **Lee-Enfield No. 4 and No. 5 Rifles** (3rd edition)

Volume 4, **Pattern 1914 and U.S. Model of 1917 Rifles** (2nd edition, revised) $16.95

The British Army's famed rifles are analyzed in detail on a part-by-part basis, complete with all inspector's and military markings.

The Mosin-Nagant Rifle (4th revised and expanded edition), by Terence W. Lapin ($19.95). A comprehensive volume covering all aspects and models from the Imperial Russian rifles to the Finnish, American, Polish, Chinese, Romanian and North Korean variations. Includes part-by-part descriptions of all makers plus all variants such as carbines and sniper rifles.

The Swedish Mauser Rifles (2nd edition), by Steve Kehaya and Joe Poyer ($19.95). A complete history of the development and use of the Swedish Mauser rifles is provided as well as a part-by-part description of each component. All 24 models are described and a complete description of the sniper rifles and their telescopic sights is included. All markings, codes, regimental and other military markings are charted and explained. A thorough and concise explanation of the Swedish Mauser rifle, both civilian and military.

A Shooter's and Collector's Guide Series

The AK-47 and AK-74 Kalashnikov Rifles and Their Variations (3rd edition, revised and expanded), by Joe Poyer ($22.95). The AK-47 and its replacement, the AK-74, are examined on a part-by-part basis to show the differences between various types of receivers, other parts, and the AK and AKM models. Also contains a detailed survey of all models of the Kalashnikov rifle from the AK-47 to the AK-108.

The M16/AR15 Rifle (3rd edition, revised and expanded), by Joe Poyer ($22.95). This 155-page, profusely illustrated, large-format book examines the development, history, and current and future use of the M16/AR15. It describes in detail all civilian AR15 rifles and takes the reader step-by-step through the process of accurizing the AR15 into an extremely accurate target rifle.

The M14-Type Rifle (3rd edition, revised and expanded), by Joe Poyer ($19.95). A study of the U.S. Army's last and short-lived .30-caliber battle rifle. Also includes the National Match M14 rifle, the M21 and M25 sniper rifles, civilian semiautomatic match rifles, receivers, parts and accessories and the Chinese M14s. A guide to custom-building a service-type rifle or a match-grade, precision rifle.

The SAFN-49 Battle Rifle, by Joe Poyer ($14.95). This detailed study of the SAFN-49 provides a part-by-part examination of the four calibers in which the rifle was made, a description of the SAFN-49 Sniper Rifle and its telescopic sights, plus maintenance, assembly/disassembly, accurizing, restoration and shooting. A new exploded view and section view are included.

Collector's Guide to Military Uniforms

Campaign Clothing: Field Uniforms of the Indian War Army

Volume 1, **1866–1871** ($12.95)

Volume 2, **1872–1886** ($16.95)

Lee A. Rutledge has produced a unique perspective on the uniforms of the Army of the United States during the late Indian War period following the Civil War. He discusses what the soldier really wore when on campaign. No white hats and yellow bandanas here.

A Guide Book to U.S. Army Dress Helmets, 1872–1904, by Mark Kasal and Don Moore ($16.95). From 1872 to 1904, the men and officers of the U.S. Army wore a fancy, plumed or spiked helmet on all dress occasions. As ubiquitous as they were in the late 19th century, they are extremely scarce today. Kasal and Moore have written a step-by-step, part-by-part analysis of both the Models 1872 and 1881 dress helmets and their history and use. Profusely illustrated with black-and-white and color photographs of actual helmets.

All of the above books can be obtained directly from North Cape Publications®, Inc., P.O. Box 1027, Tustin, CA 92781 or by calling Toll Free 1-800 745-9714. Orders may also be placed by Fax (714 832-5302) or via e-mail to ncape@ix.netcom.com. CA residents add 8.75% sales tax. Current rates are any two books: Media Mail (7-21 days) $3.95, add $0.95 for each additional book. Priority Mail: 1-2 books $6.95 or 3-10 books, $11.95. International Rates on request to E-mail address: ncape@ix.netcom.com. Visit our Internet Website at http://www.northcapepubs.com. Our complete, up-to-date book list can always be found there. All book prices and postage rates subject to change.

Note: Postage for orders placed more than one year after the copyrighted date of this book (see page ii) may have changed. The U.S. Postal service now revises rates every twelve months. If you have any questions, phone 800 745-9714 or e-mail ncape@ix.netcom.com.